LEFT BEHIND

LEFT BEHIND
RURAL ZAMBIA IN THE THIRD REPUBLIC

JEREMY GOULD

The Lembani Trust
LUSAKA

http://sites.google.com/site/lembanitrust/

First published 2010 by the Lembani Trust
Lusaka, Zambia

ISBN 978-9982-9972-4-9

Typeset in Gentium Book Basic

Lembani Trust books are distributed worldwide by the African Books Collective, Oxford.
www.africanbookscollective.com

For BJ, Mbanandi, Mbachi, Ngeza and Nzomvwa – Zambia's brighter future!

TABLE OF CONTENTS

LIST OF ILLUSTRATIONS

ACKNOWLEDGEMENTS

These studies were only possible thanks to the co-operation and assistance of many people. I would like to single out, in alphabetical order, the following: Chief Mabumba, Chief Milambo (and the 400-some households in his area who co-operated so graciously with this study), Africa Chimamba, Felina Chongola, Mike Cowen, W. Ngosa Chungu, Esko Ikävalko, Jean Jappy, Wynter Kabimba, J.K Kamuchoma, M. Kamukwamba, M.D. Kamwansa, Kari Karanko, Misael Kokwe, Shem Kombaniya, Jaakko Koskimies, Helene Maria Kyed, Liisa Laakso, Björn Lindgren, B.C. Lukonde, Ndimanye Lwenje, E. Makamba, Gun Mickels-Kokwe, Stephen Mutambalika, Gilbert Mwape, K. Mwila, Klaus Talvela, Elina Visuri, Salla Volanen, Chief Inyambo Yeta and Ben Zimba. I am also grateful to Joan Haig, Lawrence Dritsas and Marja Hinfelaar of the Lembani Trust for facilitating the publication of this collection.

LIST OF ABBREVIATIONS

ARPT	Adaptive Research Planning Team
CCS	Co-operative Credit Scheme
CRC	Constitutional Review Commission
DC	District Council
EU	European Union
FAO	Food and Agriculture Organization
FRELIMO	Frente de Libertação de Moçambique
GRZ	Government of Zambia
GTZ	Deutsche Gesellschaft für Technische Zusammenarbeit
HoC	House of Chiefs
IFAD	International Fund for Agricultural Development
IMF	International Monetary Fund
LRDP	Luapula Structural Development Programme
LCU	Luapula Co-operative Union
MMCS	Milambo Multi-purpose Co-operative Society
MMD	Movement for Multiparty Democracy
PEAR	Participatory Extension and Research
PF	Patriotic Front
PFP	Policy Framework Paper
SAP	Structural Adjustment Policy
SIDA	Swedish International Development Cooperation Agency
UBZ	United Bus Company of Zambia
UDA	United Democratic Alliance
UN	United Nations
UNIP	United National Independence Party
UPND	United Party for National Development
USAID	United States Agency for International Development
ZCBC	Zambia Consumer Buying Corporation
ZCF	Zambia Co-operative Federation

1

INTRODUCTION

Deregulation and Rural Stagnation

Good news from Africa is rare these days, but recently Zambia has been getting pretty upbeat reviews from economic analysts. The kwacha has gained in strength and macroeconomic indicators are improving. A massive debt write-off in 2005, in concert with high (if fluctuating) copper prices has encouraged a steady stream of foreign investment. After a long, dry spell of decline and disappointment, at least some Zambians seem poised to reap a richer harvest.

Economists of the neoliberal persuasion are convinced that the seeds of this apparent turn of fortune were planted in the early 1990s when the newly elected government of the Movement for Multiparty Democracy (MMD) initiated one of the most radical programs of structural adjustment hitherto seen in Africa, or indeed anywhere. Over a very short period in the early nineties, the MMD undertook to dismantle Zambia's mammoth parastatal economy, parcelling and privatizing hundreds of public companies. Adhering with emphatic zeal to the neoliberal dogmas of the International Monetary Fund (IMF), the MMD government removed barriers to foreign imports and to the expatriation by foreign investors of their windfall profits. Where the Second Republic of President Kenneth Kaunda and his United National Independence Party (UNIP) premised its policies on the need for firm state regulation of the market, the Third Republic under the MMD has dismissed the need for state stewardship of the economy, favouring a deregulationist regime. In the short term, deregulationist measures did little to modernize the stagnated economy, but they did contribute to the decimation of Zambia's fragile manufacturing sector and led to a huge loss of jobs in formal employment (Larmer 2005).

Liberalization and privatization are supposed to open up new opportunities for the business community. In reality, decades of colonial discrimination against African business, followed by three decades of post-colonial parastatal monopoly, left ill-prepared private entrepreneurs unable to respond to the possibilities posed by deregulation. Still, commerce has grown and foreign business has expanded its influence on Zambia's economy. Some effects of commercialization are quite dramatic: Lusaka especially has experienced an unprecedented real estate boom, with property values rocketing and the city itself exploding outwards along all major transport routes. Important segments of the already privileged 'line of rail', Livingstone and the Copperbelt, in addition to Lusaka, have been transformed by the market-driven policies. Visibly, an embryonic middle class is taking root.

Despite obvious improvements, it is evident that all is not well in the land. Both rural and urban poverty have reached unprecedented levels. Few other nations have a lower life expectancy at birth than Zambia, now scarcely above 40 years (World Bank 2009). Real incomes are half what they were at independence.

The hurried and poorly supervised privatization exercise allowed hundreds of millions of public dollars to slip through legal loopholes into foreign bank accounts (Craig 2000). For the most part, the beneficiaries of these dubious windfalls - ruling party politicians and foreign speculators - did not reinvest in productive industry and job creation. Indeed, the domestic manufacturing and processing sectors have yet to recover from structural adjustment. Meanwhile, vast tracts of rural Zambia have slipped inexorably back into a Neolithic state of quasi-subsistence. Already in the late 1980s, the IMF forced the UNIP government to cut back public resource transfers to rural populations in the form of agricultural subsidies. MMD executed the *coup de grâce* and abolished virtually all public support to rural livelihoods (Mickels 1997, Pletcher 2000).[1]

Market 'reforms' have pushed geographically remote Zambians towards the cities and out of the fragile markets cultivated by UNIP's rural policies and, ironically, undermined the transformative effects of an expanding money economy. Meanwhile, the same rural communities have been ravaged by TB, malaria and HIV/Aids while public health and educational services have continued to deteriorate. Rural economic withering has fur-

---

[1] For a discussion of the most significant residual of UNIP-era rural policy, the Fertilizer Support Programme, see the Epilogue to this volume.

ther catalyzed the flow of the young and ambitious to line of rail towns. Regrettably, urban job creation has not kept abreast of the real estate boom and burgeoning slums along the line of rail have become crowded pens for the unemployed and disillusioned.

The dramatic response of the urban poor to the Patriotic Front's populist rhetoric in the September 2006 elections was a clear indication that the bipolar stratification of the population into haves and have-nots is a political powder keg (see Chapter 6). The most important lesson of those elections is not about partisan politics, however, but about the economy. For all its localized successes, the prevailing market-driven policies clearly lack the means to halt progressive immiserization. Above all, deregulationist economic policies of the Third Republic have failed in forging dynamic links between agriculture, industry and trade, and in creating viable and reliable income opportunities for the mass of the population.

Could things have been done differently? Almost everyone agrees that private entrepreneurship must be stimulated in order to raise productivity and create jobs. But is the current hyper-liberalized policy regime the best, or only, possible response to a set of inevitable imperatives imposed by the global market?

The basic analysis underlying the neoliberal reforms is deeply ideological, but not patently false. For all its good intentions, the Second Republic was an economic failure. Decades of parastatal paralysis left the nation's industrial infrastructure in ruins and the government was up to its ears in debt to international financiers. While much of the blame for the dilapidated state of the late UNIP-era economy can be attributed to external shocks, the Zambian economy on the eve of the Third Republic was indisputably bankrupt; no government or financial institution on earth was prepared to salvage the decrepit and unproductive parastatal behemoth (Seshemani 1998). As a consequence, the newly elected MMD leadership had only marginal room for manoeuvre. The wave of popular support that brought the liberal reformers into power in 1991 was genuine, but fragile. MMD's first President Chiluba and his political lieutenants were acutely aware that their mandate was contingent on delivering material benefits to a massively disgruntled population; and time was of the essence.

Zambia appeared to have little hope of recuperation without far-reaching renovation of the copper mines, and the only realistic source of investment was international mining interests. Clearly something needed to be done, it was felt, to create a 'more conducive environment' for foreign investment. Following Washington's lead, the first MMD regime under Chiluba followed the policy beacon of deregulation. Economic policy aimed exclusively at minimizing state involvement in the economy would purportedly 'free up the market.' Zambia's traditional 'development partners' (now minus the Soviet Union) were fully supportive of this abrupt about face in national economic policy. Indeed, by the time the new Zambian leadership took office, policies based on the so-called 'Washington Consensus', i.e., market liberalization and privatization, were an inescapable precondition for international financial support (Craig 2000).

The Politics of Deregulation

Structural adjustment, then, was probably unavoidable; but the form it took, the specific policies that were prioritized and the way they were implemented, was strongly shaped by political contingency. UNIP had fostered its political power base through subsidies and other forms of patronage; when money to reward people's loyalty ran out, President Kaunda's reign crumbled. MMD inherited an empty treasury and was obliged to forge a different approach to politics.

Despite its de facto monopoly of power, the new MMD government could not even dream of establishing the sort of country-wide grassroots party-state machinery that UNIP had achieved at its peak (see Chapter 1). Out of necessity, the MMD set out to redefine the nature of the state, and the relationship between people and government. Where the Humanist ideology of UNIP envisioned the state as an extrapolated extended family with President Kaunda - the Father of the Nation - at its head, MMD's political message was of individualism, self-reliance and sacrifice. In practice, the MMD turned its back on the rural clients of UNIP's patronage and focused on fostering the support of its foreign backers and urban-based business interests. Unfortunately, by this time Zambia's business infrastructure was in shambles and the market failed to respond to hasty efforts at jump-starting the economy. In consequence, the MMD leadership resorted to increasingly questionable means to stay afloat (Lewis 2009).

In retrospect, it is evident that the deregulationist reforms had clear political objectives, for both foreign institutions and for their domestic clients. Indeed, the specific policies were more successful in meeting short-term political goals than they were in modernizing the economy. Zambia's international partners were primarily concerned with demolishing the parastatal industrial sector and thus opening up the economy to foreign goods and capital. Irrespective of professed intentions, the ultimate outcome was to allow for the re-appropriation of Zambia's copper reserves and the mining infrastructure by international capital.

For domestic political actors, economic liberalization implied windfall benefits. Financial deregulation, in particular, allowed new opportunities for speculative investment in rapid turnaround, high-profit import-export ventures. Liberalization also eased the expatriation of kwacha assets to overseas bank accounts. Undeniably, deregulation transformed urban society. It clearly fosters middle class formation, a necessary if insufficient precondition for economic modernization, and has encouraged the professional classes - epitomized by an emerging corps of affluent lawyers, accountants and managers - to entrench their entitlements in the new dispensation.

Deregulation has, then, two victims: one is the developmental state, and the concomitant idea of the common good. The wanton destruction of the parastatal economy created an economic and political ethos of 'every man for himself.' The public economy associated with the postcolonial state became fair prey for unscrupulous plunder. Zambia's urban classes are apparently ambiguous about the effects of this ethos. On the one hand, the throngs of sharply-attired, wine-sipping urbanites that patronize the panoply of fancy bars, hotels and restaurants sprouting up along the line of rail clearly revel in their new-found affluence. On the other hand, they bemoan the lack of state leadership, as evidenced by the strong negative reaction by civil society to President Chiluba's acquittal of corruption charges in August 2009.

Incredible effort and energy has been invested in mopping up the kleptocratic excesses of deregulation in the Chiluba era, albeit to little tangible effect. Meanwhile, the other primary casualty of economic deregulation, the millions of economically disenfranchised men and women trapped in stagnating rural communities, continue to have negligible currency in political life and economic policy.

The redefinition of rural Zambia as redundant to 'development' renders villagers lacking the means for commercial production bereft of economic agency and citizenship. The deregulationsist policy regime entails no official recognition of their economic rights, or of their creative powers. The broad stratum of 'emergent' farmers that had been the express target of the Lima policies of the seventies and eighties are now deemed irrelevant for the modern economy. At best, they are potential clients for political patronage in election years. Prevailing development rhetoric recognizes the need for 'poverty reduction.' And yet, official poverty reduction strategies remain captive to the ideology of market-driven development, and thus have little to say about the crucial role of the state in enhancing agricultural productivity and fostering rural livelihoods. This developmentally 'irrelevant' section of the Zambian population constitutes anywhere from half to four-fifths of the nation.

A Long-Term Perspective

The studies in this book were written between 1990 and 2005. The studies are primarily concerned with the southern plateau of Luapula Province, one of the least developed areas of the country. Some of the empirical data is dated but yet, despite their evident limitations, these studies bring out a number of important points that need to be made today as much as they did a decade ago.

The central lesson is that farmers respond creatively to crisis. They cannot afford to succumb to defeat and are incessantly developing new ways of dealing with adversity. Often, this involves innovative changes to their systems of production, to land and labour management, crop marketing and stop-gap financing. There are thus always forms of innovation and momentum at play in the rural economy that can provide a productive entry point for development policy. Even considering the dire straits of Zambia's public economy in the Third Republic one could, and still can, envision policy measures that would support farmers' own efforts to develop rural livelihoods. Even minor investments can make a decisive difference, but they must be intelligently targeted.

Rural Zambians are dependent on some form of public support in order for them to enjoy the basic rights of economic citizenship; the market alone will not solve their problems. Policy measures that build on the creative innovations of rural women and men do not coexist comfortably with the grandiose rhetoric of neoliberal deregulation. Rural economies are complex and fragile; they cannot be targeted with vague ideological commitments to 'market liberalization' or 'poverty reduction.'

Finding ways of supporting the efforts of rural producers requires careful study; above all it requires the time and tenacity to spend time with farmers in their villages and fields, to listen to their own analyses, and to help them formulate strategic options based on a sober understanding of the real constraints imposed by national and international economic structures. Local producers will often know what needs to be done, but the means for them to articulate these needs in a politically effective way are largely lacking.

The basic premise connecting the different analyses is the need to understand the economics and politics of rural communities in order to identify policy alternatives. Mine is certainly not the only such work on rural Zambian society in the Third Republic; yet, too little of such empirical analysis is available to the Zambian public, to politicians, to policy makers and to my fellow researchers. When there is little public knowledge of the dynamics of rural society it is easy to ignore the real development policy options available. It is my hope that this slim volume will help stimulate policy debate and further research on the Zambian rural economy.

The second chapter reproduces an essay originally published in *The Post* newspaper in 2003 concerning the roots of rural disenfranchisement in the Third Republic. Together with the final essay - a brief analysis of the 2006 elections - these pieces attempt to flesh out the historical and political context through which the earlier studies in-between might be approached. These are the most general parts of the book and may not contain much new information for a Zambian readership. Other readers will hopefully find the contextualization useful.

The third chapter is an overview of the state of the Luapula economy on the eve of market liberalization. Written in 1991, just prior to the elections that replaced UNIP with MMD, it offers an analysis of the actors and structures then in place, and sketches possible scenarios which could unfold with the coming of deregulationist policies. Nothing has been altered in the original text (other than the title) and I invite the reader to reflect on why this analysis was dismissed by the commissioning donor agency, and on what might have happened had it been taken seriously.

The fourth chapter, the longest essay in the volume, offers an in-depth analysis of the impact of market liberalization atthe household level in one rural community. It is based on a comparative analysis of original survey data collected in one rural chiefdom in 1987 (at the zenith of the Lima policy campaign) and again in 1996.

A relatively recent study, which comprises the fifth chapter, addresses the political dynamics of rural Luapula in the Third Republic by focusing on the changing role of traditional authority in Zambia's fledgling democracy. More than the other essays, this piece attempts to link grassroots political changes with transformations in the political system at the national level.

In chapter six is found is an analysis of the national elections in 2006. Following this, an epilogue, written for this book, attempts to bring the collection up to date with a brief commentary on the current state of rural development policy, and its roots in a culture of elite entitlement.

Together with my earlier books *Luapula. Dependence or development?* (1989) and *Localizing modernity. Actors, interests and associations in rural Zambia* (1997), the current volume, *Left Behind*, forms the final instalment in a 'Luapula trilogy,' dedicated to documenting and analyzing the political economy of rural change in one corner of Zambia. In bringing this project of twenty-five years to a close, I have two heartfelt hopes: one, that this work will in some small measure reciprocate the generosity and warmth my family and I experienced during our time in Luapula; and second, that these three texts will stimulate pride among all Zambians in their rural roots, as well as clear-headed reflection on how disenfranchised rural communities might be once again genuinely included in the nation.

2

BEYOND THE GOVERNANCE CRISIS
(2003)

Zambia has earned a reputation for pioneering political progress in Africa. The peaceful transition to multi-party politics in 1991 and the people's univocal rejection of the unconstitutional third-term bid in 2001 have provided inspiring precedents for progressive actors in neighbouring countries and across the continent. The current administration's campaign against corruption, and the associated lifting of the defeated President's judicial immunity, while somewhat more controversial, have also commanded respect.

True, Zambia's productive infrastructure is in ruins, social conditions desperate and abject poverty the norm. Yet Zambians can justifiably feel proud of their political achievements. Zambia is clearly the one African country where civil society has matured most rapidly, and where a broad coalition of progressive social forces has asserted itself successfully on behalf of greater political accountability in a focussed and organized fashion.

For this very reason, a great many Zambians find themselves stunned with disbelief at the extent to which the political situation has deteriorated onto the verge of chaos within a few short months. As power splinters, leadership evaporates, and the once-banished scourge of ethnic populism raises its hoary head, Zambians are beginning to wonder whether the peace and stability that has graced their country will survive until the 40 years' independence celebrations next year.

The Roots of the Crisis

It is difficult to pinpoint the precise roots of the current crisis. I will argue below that Zambia's political system never fully adjusted to the multi-party dispensation of the 1991 Constitutional reform. With equal cause, one could argue that we are still reaping the bleak harvest of the one-party state introduced in 1972. Many would like to go even further back and blame colonial shenanigans for present difficulties.

Whatever benchmark you chose, it is evident that Zambia has been crippled time and again by a system of governance which can only function if the omnipotent Executive has the will and the skill to put the common good of the nation above his (or her) personal interests. This is the model of governance that the Father of the Nation, Kenneth Kaunda, created for himself, and which he wielded with greater and lesser success for 27 years. Kaunda's vision of governance was predicated on a single, centralized site of political, moral and legal authority commanded judiciously by an all-powerful executive – absolute power tempered with moderation and re-straint. Kaunda was (and is) a human being with all the attendant human frailties, yet it is clear in retrospect that his rule was guided by commitment to a sophisticated bundle of political virtues. So much so that, in the end, Kaunda was prepared to step aside and allow power to pass on to his opponents when it became evident that he no longer enjoyed the full allegiance of the Zambian people.

The 1991 revision of the republican constitution that re-allowed multi-party politics in Zambia did nothing to alter the status of the executive, nor change the model of governance. Despite the new rhetorical commitment to liberal democratic norms, President Frederick Chiluba was allowed, even obliged perhaps, to perform the duties of the executive as if his political, legal and moral authority was above dispute. It would seem – again in retrospect – that Mr Chiluba's personal and ethical qualities were not commensurate with the responsibilities of such powers. Whether due to weakness, design or frustration, his administration deteriorated into a state of kleptocracy. That is, the personal interests of the executive eclipsed his stewardship of the common good. Chiluba's perversion of power damaged the political system gravely, and it will take a painfully long time to repair the economic, political and moral injury. But while the nation fixates on its disappointment, shame and lust for retribution an important lesson is being overlooked. The deficiencies of the Chiluba administration cannot be blamed solely on weak or evil politicians. The MMD kleptocracy was not simply the product of corrupt individuals; it was the direct outcome of a structural mismatch between the prevailing system of governance and the new dispensation of power in the polity.

The Structure of Power

Let me explain. Under the Second Republic, power was concentrated in Kaunda's hands, and those hands reigned over a comprehensive UNIP organizational structure that extended down to the grassroots of Zambian society. Every ward, section and branch had its Party officials; in the late 1980s, when I carried out research in rural Luapula, UNIP was a pervasive presence wherever one went. Despite its landslide electoral victory in 1991, the Movement for Multiparty Democracy has never achieved this level of hegemony. Even had it been financially feasible (and the donor-driven privatisation of the parastatal sector made it impossible to maintain the patronage structures of the Second Republic), the ethos of multiparty politics was hardly compatible with establishing a grassroots organizational infrastructure after the UNIP model. Cut off from state resources, the UNIP network shrivelled, and nothing took its place.

Almost overnight a situation emerged in which no regulatory structures linked to the centralized authority of the executive existed at the grassroots. This sudden regulatory vacuum at the base of society upset a long-established political equilibrium. The ambitions of various players – for power, influence, and the control of resources – had been held in check, to a large extent, by the patrimonial structures of the UNIP machinery. As the regulatory mechanisms of the Party-State crumbled, new self-interested players with a wide range of interests and ambitions have begun to emerge. Some of these are associated with the rapidly changing gallery of political parties, some with business and commercial interests. Others are linked to the mushrooming array of organizations that compete for government and donor contracts. Traditional leaders have also found new space to assert their primordial claims to local authority. With the lack of effective co-ordination and control, power – especially at the base of society – has become fragmented and diffuse.

'One-Party Participatory Democracy' Revisited

Now, it may well be that the image we have of the authoritarian, top-down, centralized Party-State apparatus of the Second Republic needs to be re-evaluated. The UNIP government, one will recall, claimed to formulate and implement its policies in the spirit of 'one-party participatory democracy.' As the economy deteriorated beyond repair in the 1980s, and people tired of Kaunda's excuses for the constant shortages of basic commodities, the rhetoric of participation and democracy was evoked with increasing cynicism. Yet looking back, perhaps the extent to which the pervasive UNIP party apparatus exercised control on the Executive was not negligible. One shouldn't forget that the exercise of governance by Kaunda and UNIP was enmeshed in a routinized schedule of annual Party conferences at the District, Provincial and National levels. The upstream Party leadership was regularly exposed to the rank and file of grassroots cadres who could, with little ambiguity, hold the President and the Central Committee responsible for any resources or development that accrued to their constituencies. To some extent, the massive and nearly countrywide rejection of UNIP at the 1991 polls was a result of the transparency of its failure to deliver on its promises.

What needs to be understood is that the subtle relationship of reciprocity between the ruler and the ruled that under-girded the politics of the Second Republic, and which provided a degree of legitimacy for nearly twenty years to the UNIP Party-State, is irrevocably gone. Even were Kaunda to enjoy a second coming, the relationship he had with the Zambian people can never be restored. The character and dispensation of power in Zambian society has changed too much. And yet, both Kaunda's successors have sought to rule as if they could take the centralized authority of those foregone days for granted. This is the mistake that led Mr Chiluba to lose touch with reality; it is similarly the source of Mr Mwanawasa's growing frustration.

This lesson is especially timely today, as Zambia prepares itself, yet again, to reconsider the parameters of its Supreme Law, the Republican Constitution. One should not make the mistake of viewing the deepening crisis of the Mwanawasa administration as simply the result of a flawed election, the personal qualities of the executive, or an internally divisive governing party. The fundamental problem is structural, and it can only be resolved by a profound rethinking of the governance relationship – the complex bond between the ruler and the ruled. Zambia can only be ruled effectively and equitably when the system of governance is adjusted to contemporary realities. At the core of this is the constitutionally enshrined relationship between the Executive and the People. As long as the system of governance is based on the premise that whoever sits in State House is the unquestionable source of political, legal and moral authority, Zambia will wallow in a mire of endemic political crisis.

The current debate about the democratic reform of the Republican Constitution provides an invaluable opportunity for the profound rethinking required. A good constitution, while no doubt necessary for resolving the deep-rooted crisis in the political system, is nonetheless not in itself sufficient. The main issue is not the formal model of governance, but the real-life relations of governance that need to be re-established between State and Society. The Constitution can only lay the foundations for reciprocal and hence deeply democratic relations of governance. The decisive factor will always be the style of governance and the skill of the leadership in establishing an affective basis for trust and loyalty.

This is why the process and procedures by which the new Constitution will be written, adopted and enacted are so important for the future of democratic governance in Zambia. The symbolic act of allowing the Zambian people, through their chosen representatives to a Constituent Assembly, to actually write their own Constitution could go a long way to creating a fund of social capital that the current administration, and the nation as a whole, desperately needs to survive these trying times.

3

WAITING FOR CHANGE
Structural Adjustment & Luapula's strategic options
(1991)

Preamble

Within the context of its chronic battle against world market marginalization and deepening domestic recession, Zambia has entered yet another
period of economic transition. Impeti for the current changes come from
two disparate directions. On the one hand the World Bank/IMF power block
(behind which most bilateral donors rally) demands an extensive package
of economic reforms. On the other hand, major interest groups in Zambian
society (commerce, wage employees, intellectuals) can no longer countenance the costs of current economic policies and are seeking political
means to instigate economic changes. These political pressures in concert
with an economic situation which can justifiably be characterized as 'out of
control' has led the government to reiterate its commitment – abandoned
once in 1987 – to an IMF-style Structural Adjustment Policy (SAP) program.

The objectives of a SAP are in brief:

> to restructure and raise the efficiency of the economy through active exchange
> rate and monetary policies; price control and phasing out of Government
> subsidies; elimination of market and price distortions; parastatal reforms and,
> generally, through introduction of a market-oriented economic system which
> promotes private initiatives and competition as a means of improving econom-
> ic efficiency (LRDP n.d.: 33).

It should be apparent that these measures imply profound social, economic
and political consequences for Zambian society. Economic historians generally agree that competition, unbridled market forces and commercialization correlate with increased social stratification and growing
discrepancies in the distribution of resources among classes. One outcome

of the ongoing transition is a radical reduction of the role of the public sector in the Zambian economy. A parallel trend toward improved public sector efficiency has already instigated the divorce of the political apparatus of the Zambian state (UNIP's single-party regime) from the administrative organs of government. Clearly the ongoing transition – whether economically successful or not – will have profound social and political consequences.

Because of all of this, the policy environment within which the Finnida-funded Luapula Rural Development Program (LRDP) operates is obviously modulating into something new. What the new environment will comprise is not clear. Even were the parameters of change clear today – which they certainly are not – administrative renewal in Luapula's rural outposts could not be expected to generate enough immediate pressure on the Program to alter its operations for quite some time. What tentative guidelines might now be compiled about the implications of these trends for forward planning of the Finnida program?

Given the increased uncertainty coefficient attached to all factors affecting 'development' in general and donor interventions in particular, the best counsel at this point may be to lie low, maintain current operations while carefully observing the tendencies and magnitude of the transition. This 'passive' strategy would (only) cost about 17 million Finnish *markka* per annum (the drain on Finland's balance of payments is of course considerably less) and would postpone the eventual need for a period of intensive activity (assessment, planning, revising management and administrative monitoring systems to correspond to a new structure of operations) and new administrative costs for an indefinite period of time. Judged by technocratic criteria alone, a strategy of alert inactivity would appear most appealing.

Yet even the maintenance of current operations cannot be deemed a neutral action. The prevailing structure of Finnida investment in Luapula reflects economic assumptions and political realities which events of the past two years (since 1989) have rendered archaic. Continuing to plough voluminous financial resources into institutions and activities which may not provide momentum for constructive change can have a demoralizing and debilitating effect on the economy and its key actors. On the contrary, a more aggressive, positive approach could conceivably help sculpt the landscape of the region's future. As the main financial force in Luapula,

Finnida can be said to have special responsibilities to shoulder, even in times of uncertainty and change.

Toward the mid-1980s Zambia's public debt became unmanageable in the eyes of Zambia's major financial watchdogs, the IMF and the World Bank. Further financial assistance was made contingent on SAP-type economic reforms, among them the significant 'liberalization' of trade in primary agricultural commodities. Zambia balked in 1987 but finally acquiesced to IMF provisions in 1989. Since 1990 Zambia has been involved in the process of freeing the grain market and of deconstructing the intricate system of overt and covert subsidies built into the prices of agricultural services and commodities.

The processes instigated by these measures have several implications of relevance to the Finnida program. In brief, these might be elaborated as follows:

a) There has begun a vital resurgence of private initiative in agro-commerce (marketing, transport, trading, processing) on all levels (ward/community, district, intraprovincial, inter-provincial). Private operators have been active in this sector since colonial times, yet public discouragement has curtailed the development of the entrepreneurial class. Now this appears to be changing with private businessmen rising into the forefront of the economy.

b) Market-oriented primary agricultural production is diversifying away from maize. Instead of maize, greater volumes of 'traditional' food crops such as cassava, groundnuts and beans are finding their way onto the local and regional market. Alongside these, the marketing of some non-traditional cash crops (rice, vegetables) is also on the rise.

c) As subsidized marketing services (provided through primary co-operative societies) dissolve the importance of local markets is likely to grow. The study data indicate that this is clearly true for the Provincial core areas (Mansa, Kashikishi), but there appears to be a similar tendency for more intense commodity exchange in lower level population centres as well.

d) As already noted, a radical change can be seen in the role of government from that of a monopolistic provider of services and guardian of commodity flows to one of a limited facilitator in limited areas. In other words, public sector institutions are being redefined as simply one set of actors among many.

Finnida has traditionally collaborated with the government and channelled aid into public administration and parastatal or heavily subsidized economic institutions. New forms of 'co-operation' with new kinds of 'partners' will be necessary in the unfolding situation. It will thus be difficult – and unnecessary – to restrict the rhetoric of situational analyses to the accustomed framework of the 'official' discursive practices which have tended to reflect a narrow governmental/diplomatic etiquette. Taking an active role in the new situation will require an ability to empathize with the needs and mentalities of a wide range of actors beyond including those outside familiar ministerial and parastatal corridors. The elaboration of a new strategy of assistance will require new perspectives, new tools of analysis and a new language (or languages) for discussion and negotiation.

The crucial task of situational analysis and policy discourse at this stage is to attempt to redefine the respective roles of the public and private sectors – and of specific actors within each sphere – in the development of the economy. Questions that must be asked include: What activities and what actors can be expected to play a constructive role in 'development' during the next decade, and under what conditions? What are the 'comparative advantages' of different actors? What political contingencies prevail? These questions have strong ideological connotations. To answer such questions fruitfully it is necessary to examine briefly the ideological milieu within which policy discourse is being carried out.

The Ideological Environment

A Disclaimer

The following remarks on the ideological context of development interventions in Luapula are overtly provocative. The intention, however, is not to discredit the efforts or motives of any particular actors who have participated in the development of Luapula. It is accepted that everyone is a captive of the intellectual milieu of the time. Ideology is not only unavoidable; it has a functional aspect in certain organizational settings. Within complex bureaucracies, like GRZ (Government of the Republic of Zambia) or Finnida, ideologies provide the necessary justification for choosing between equally uncertain alternatives, and thus prevent work from grinding to a stop.

Still, ideological frameworks periodically develop anachronistic features as the real-world environment changes and these changes render the premises upon which the ideologies rest transparently false. At such times it is valuable to bring the discrepancies between ideology and reality out into the open. The ongoing economic and political changes in Zambia have created such a situation. While the observations below may appear extreme, they can hopefully help stimulate a less encumbered discussion concerning alternative paths of action.

The Idea of Development

Donor aid-policy thinking appears to be pervaded by pessimism. The development optimism and trusting atmosphere which prevailed from Zambia's independence well into the 1970s has dissolved into a new 'realism' according to which hopes for greater welfare and development have been replaced by prescriptions for self-restraint, austerity and 'adjustment'. Call it cynicism, aid fatigue, avarice or whatever, the belief among aid authorities that 'development' in Zambia/Africa is imminent or even possible stands today on shaky legs.

In Finland, where the relative growth of the economy has slowed down to 'recession' levels, development aid funds have been slashed with impunity. The few voices raised in protest have been content to defend aid funding solely for social sector allocations (water, health, food aid for catastrophe victims, social infrastructure).[1] Humanitarian considerations continue to provide some justification for aid spending, but the idea of development meaning *a structural transformation of the economy as a whole* seems to have disappeared. Indeed, within European political discourse 'development' has been diluted into a synonym for 'survival'.

From the Zambian end the perspective is somewhat different. Structural adjustment policies have been welcomed by political progressives and reformist forces (despite the associated austerity measures) because they have been seen as a means of decapacitating the system of economic and political patronage which obstructed economic renewal under the Second Republic. When reformists cite market liberalization and unbridled entrepreneurship as the prime vehicles for development they signify the need to dislodge the (United National Independence) Party and its government from its monopoly of control over resource flows. Conservative forces – which remain firmly established in leading positions of the government –

---

[1] e.g, a letter to the Editors in *Helsingin Sanomat* (18 August 1991); *Uudistetun kehitysyhteistyön puolesta* ('For renewed development co-operation' – a statement by the Board of the Finnish NGO Kepa (6 June 1991).

are caught between the liberalism of the reformists and the cynicism of the donors while trying to devise a constructive policy offensive that could secure their political futures.

The dismal failures of the Zambian economy under the Second Republic have undermined confidence in its official structures and institutions. This scepticism about public sector potential unites donors and reformists. Even the conservatives have acknowledged a need to streamline public sector economics, yet they appear constrained by their own organizational inertia into merely defending the status quo, thus weakening their credibility with foreign financiers. Lest these divisions appear too clear-cut, it should also be held in mind that donors have dealt with GRZ and its representatives, planning, implementing and evaluating programs revolving around public sector investment for a quarter of a century. Most bilateral donors, like Finnida, are themselves public sector organizations and are staffed by people who share a generic civil servant suspicion of commercial actors.

A 'market-centred' development strategy must come to terms with these realities. Under Luapulan conditions 'liberalization' reveals itself as a very problematic catchword, certainly not as a panacea for all the ills of an economy in advanced stages of disintegration. Still, it cannot be argued that economic reforms would not be urgently needed, or that the past recipe of over-subsidized metastrategies, bureaucratic behemoths and parastatal straightjackets has outlived its usefulness. But as several commentators have already noted, the bankruptcy of the Second Republic only created sufficient *political* space for the SAP. In strictly *economic* terms there are no factors intrinsic to the current situation which dictate the kind of complete about-face in policy which the SAP -type reforms imply. The SAP program reflects choices made in a situation almost deplete of any constructive political, economic or ideological alternatives. Awareness of this explains the contradictory nature of the ongoing transition. The SAP reforms aim at devolving economic controls from the state to 'free-market forces'. In many respects, however, the process of this devolution would require that the reigns of the economy (with respect, for example, to foreign exchange markets) be retained in the firm hands of the state.

This paradoxical yearning for both liberalization and central control is an honest reflection of the contradictory pressures with which Zambians must cope. Any program of development interventions, such as the Finnida LRDP cannot avoid getting caught up in the same equivocations. Hence, on an ideological (or 'theoretical') level there is broad acceptance among donors of the need to stimulate the 'private sector'. Yet, for lack of clear alternatives – precisely because the economic domination of the state has inhibited the private enterprise for so long – external agencies recurrently find themselves dependent upon governmental or parastatal bodies for the delivery of support to the 'private sector'. This ambivalence toward liberalization might thus be seen as a natural adaptation to the structural contradictions inherent in the breakdown of one system and the genesis of another.

These ideological contradictions are very much on the surface as reformists and conservatives prepare to measure their power bases in the multiparty elections scheduled for October 1991. But the underlying economic problems extend beyond the scope of the current political campaign. For the purposes of this essay, these problems translate into the dilemma of elaborating a realistic and operational conception of 'development' for Luapula within the scope of the prevailing political and ideological parameters.

For the time being, however, the economy continues to be dominated by the state and regardless of the velocity and extent to which liberalization proceeds in the foreseeable future, important economic functions (e.g. social infrastructure) will remain within the public domain. The new order cannot entail a clean break with the past nor, consequently, can policy choices fall along clear-cut lines, such as public vs. private sector investment. This suggests that concrete decisions concerning donor investments will be more complex than was usually the case under pre-SAP inter-governmental bilateral co-operation.

Choices will be difficult and it will not be possible to know with certainty where any given decision will lead. What is hopefully possible is for all parties to require that decisions are informed by a broadly-based debate around alternative paths of action. The Finnida program, its personnel and its implementing agency are important actors in the political sphere in Luapula. An espoused commitment to the democratization of society commends the Program to respond responsibly to the realities of this role and to take initiative in maintaining such a policy discourse.

The Finnida Program[2]
The substantial relative weight of the Finnida program in the region makes

---

[2] 'Finnida' was the name use by the Development Co-operation department of the Foreign Ministry of Finland from about 1975 to 1991. Since then, Finnish aid has been branded as a product of the Ministry of Foreign Affairs of Finland.

it vital that the development vision behind its planning and implementation be as explicit and well-conceived as possible without becoming inflexible. Sociologically speaking one of the greatest dangers is to inadvertently attempt to 'freeze' the local social structure in conformance with moral guide-lines culled from European/Nordic experience. Development objectives seem often to be justified on the basis of an intuitive conception of what should constitute a 'just' distribution of resources, or what should be the 'natural' role for rural Africa in the national economy, seen from a Scandinavian perspective.

Thus, while it may seem socially moral to privilege the 'poorest of the poor' (e.g. single-female headed households) for program inputs, there is little historical or theoretical justification for thinking that such an allocation pattern will lead to 'development' on a larger scale, or beyond the scope of project duration. This is not to say that aid should not contribute to short-term humanitarian goals, such as assisting in the provision of social services when necessary government resources cannot be mustered from domestic revenues. The point is that the *fundamental* objective of 'development' aid should be to enhance the capacity of the economy to generate adequate domestic revenues for the sustainable provision of basic services such as clean water, health facilities, education, and so on. Humanitarian aid and 'development' aid can and often should complement one another, but they are not synonymous. Nor should humanitarian criteria be substituted for the analysis of long-term development considerations in program design (unless we truly believe that donor agencies are committed to provisioning the basic needs of the 'poorest of the poor' *ad infinitum*).

The ideological underpinnings of the Finnida program are also reflected in its equation of agricultural modernization with rural development. It is a fact that most Luapulans are involved to some extent with crop cultivation, but it is also true that in terms of their business strategies for cash income generation a large proportion, very probably the majority of the population, are oriented toward other activities (fishing, commerce, wage employment, artisanal trades). Why have LRDP investments singled out export agriculture (hybrid maize) for their main thrust? Is it because program designers have implicitly accepted the prevailing view (*pace* urban-based Zambian planners and politicians) that the economic role of rural 'peasants' is to provide primary commodities and raw materials for the urban sector of the economy?

Why has this perspective appeared so 'natural'? In reality Luapula is a net exporter of hydro-electricity, has unique reserves of semi-tropical hardwoods, vast supplies of locally processable foodstuffs such as fish and fruit, and a large surplus population with decades of industrial labour experience – all of which could provide a basis for proto-industrial development in the longer term. With the benefit of hindsight, would it not have been equally or more rational from a 'developmental' perspective to spend the time and resources hitherto invested in the Finnida Luapula program (roughly one decade and funds in excess of FIM100 million?) to promote, say, local manufactures (including the development of suitable technologies and forms of industrial enterprise) rather than hybrid maize?[3]

The Role of Agriculture

The perspective adopted in the above remarks implies that agricultural modernization can have but a limited long-term impact on the overall development of the Luapula region. This position contradicts much of the accepted wisdom which has been explicit in GRZ development policy from the TNDP onward; that is, that the Zambian economy must diversify away from mineral extraction into agriculture, even in such agriculturally marginal areas as Luapula. The thrust of the SAP reforms are based on a similar assumption. World Bank policy papers also concur, emphasizing the need to identify comparative advantages of the production of suitable crops in each region of the country and to base agricultural development policy on such analyses. Since we are not aware of any data or analyses which would provide an *economic* justification (in a free-market context) for substantive investments in agricultural modernization in Luapula, it is necessary to examine the relationship between the ideology of agricultural development as expressed by SAP policies and the prevailing realities upon which investment decisions must be based at this stage.[4]

Channelling support for agricultural development into the extension wing of the Ministry of Agriculture, and specifically into hybrid maize promotion expresses confidence in the out-lived 'meta-strategy' of maintaining monopolistic government control over food production and marketing on a national level.

---

[3] It is probably wise to stipulate in this connection that a development policy promoting manufacturers is in no way synonymous with an investment program based on the importation of Finnish or other Western technologies. Technological (and organisational) progress can also be built upon local expertise, although this option can naturally take longer before initial results are visible.

[4] Social and political justifications are naturally another matter.

The instruments associated with the government's maize policy can be called a meta-strategy because they constitute a complex web of interventions pursuing many objectives with numerous potential repercussions. The government's maize policy catered to the political demands of its urban supporters for cheap food. At the same time, parallel instruments sought to link the fortunes of rural peasant producers to urban food needs. It was hoped that peasant participation in the national grain market would not only ensure urban food security but also enhance rural incomes and thus promote 'rural development'. By virtue of the heavily subsidized provision of extension and marketing services maize production in Luapula increased many-fold over the course of the 1980s. Nevertheless, Luapula's soil conditions are far from satisfactory for hybrid maize production and on-farm yields have remained far below commercial standards. As subsidized inputs and services have been withdrawn, production levels have begun to fall drastically.

For the better part of a decade, the Finnida program has also provided massive financial and technical support to Luapula Co-operative Union (LCU). Alongside its maize extension offence, the so-called co-operative movement has constituted the second, crucial leg of the pre-SAP food sector meta-strategy. From the first months of Zambia's independence co-operatives have been seen as the leading vehicle for the pursuit of development aims in rural areas. The GRZ 'co-operative development' strategy has entailed overloading co-operative institutions at all levels (primary, secondary, tertiary) with functions and expectations, while at the same time subjecting these same institutions to the strict controls and constraints dictated by the contradictory aims of the food sector meta-strategy. Provincial Co-operative Unions like LCU were midwifed into existence through intensive donor support. The trials of consolidating a viable co-operative marketing system in Zambia's inconsistent and often hostile policy environment have proved overwhelming. Not only has LCU persistently failed to stand up on its own feet, it has tended to cultivate exploitative and asymmetrical ties of dependency with the producers' primary level co-operative societies.

Against the backdrop of a rapidly declining GNP, the government has proved unable to stop-gap the discrepancy between high rural producer maize prices and low urban consumer mealie-meal prices on the basis of domestic revenues. Temporary solutions to foreign reserve deficits were sought through loans and spontaneous money creation. The variegated package of subsidies through which the food sector metastrategy was maintained thus implied an inflationary economy as well as a heavy supplement to the already staggering national debt.

Within the context of SAP policy, agriculture is primarily seen as a substitute for copper in Zambia's export economy. 'Agricultural development' and 'comparative advantage' are concepts which have as their point of reference the international market for agricultural commodities. For some regions of Zambia, export oriented agriculture has demonstrated commercial prospects under liberalized market conditions. Yet, for Luapula this is hardly the case. With the exception, perhaps, of small peri-urban pockets, formidable marketing costs frustrate the development of smallholder agriculture on a commercial scale. A plantation-based system would appear more promising on paper, yet the practical experiences with Kawambwa Tea and various midget coffee schemes over the past two decades have not been encouraging.

Semi-subsistence agriculture (the combination of production-for-use with subsidiary production for the market) is a major economic activity in the region. In the past research and experimentation on improving varieties, farming systems and technologies for the region's major traditional crops has been virtually ignored. Recent years have seen a shift of official interest toward these traditional food-crops, although the scope of activities has remained rather modest (a pilot-scale Adaptive Research program). It would be misplaced optimism to project adequate international demand for Luapula's traditional food crops (tubers, legumes, squashes and leafy vegetables) at economic prices. There is, on the other hand, a sizeable if limited local market for foodstuffs, both within the Province and in the relatively nearby urban centres of the Zambian and Zaïrean [Congolese] Copperbelts. These markets are limited in the sense that they are easily saturated when harvests in adjacent areas are good. Yet, with the improvement of storage facilities, there should be a basis for the considerable expansion of current production. The marketing of this surplus could also have a broader developmental aspect by laying a commercial foundation for the consolidation of the local transport and trading sectors. Thus, it can be argued that agricultural modernization in Luapula, if understood as technology-intensive production for the world market, is an unrealistic dream. On the other hand, regional development in economic and social terms requires investment in the improvement of local agricultural systems.

This is the proper context for clarifying the status of women as a specific target group for development interventions. There is by now general awareness that women play a decisive role in the management of local agricultural systems. More often than not a 'farmer' in Luapula is a woman. 'The improvement of local agricultural systems' thus logically entails involving women and supporting to a large extent the work that women are doing. If 'the improvement of local agricultural systems' is part of a broader strategy to restructure the regional economy, it follows that interventions aimed at supporting women cultivators are 'developmental', not 'humanitarian'. But women also play important non-agricultural roles in the economy. In Luapula, women have long been known for their success in trade and other forms of private entrepreneurship. It is also gender bias to single out only one of women's many roles as the crux of women in development.

These critical remarks on ideology have sought to explicate the social context of the current phase of decision-making within the Finnida program of development interventions. Seeking to make outmoded ideological structures transparent does not imply that we are now in a position to undertake a 'non-ideological' situational analysis. Quite the contrary. The ideological configuration surrounding structural adjustment is very dense indeed and the limits which concepts like liberalization, the free market, competition, public sector efficiency, rational resource allocation, entrepreneurship and private initiative place on our ability to perceive available options clearly are not visible at the present. It is in this spirit of pragmatic uncertainty, then, that the following discussion attempts to outline the main parameters of the 'development' process in Luapula and the role of the transport sector therein, by way of a close case study. The discussion is organized in the form of an inventory of the social structures through which development should occur, and of the economic actors upon whose action the content of development (and the success of development interventions) to a considerable extent depends.

Actors and Structures

The Market

Prevailing currents in Zambian development policy highlight the market. The structure of commerce in Luapula reflects a long history of jerky, quasi-cyclical development. Periods of high and consistent demand for specific commodities (including labour power) have abruptly plummeted into subsequent stages of near stagnation, only to rise again after a time. In most instances, these rapid shifts have been the direct result of inauspicious interventions by colonial or post-independence governments. Luapula's proximity to Zaire has also had a strong, if complex impact on the development of the market situation. Because of the long history of cyclicality and associated uncertainty, producers (and middlemen) have been rather responsive to short-term market signals.[5] On the other hand, these same risk factors have cultivated a general reluctance among economic actors to commit their resources fully to any one sphere of activity. Luapulan entrepreneurs involved in agriculture, fishing or commerce are polystrategic. Resources remain mobile and are rotated between various ventures. In corollary, concentrated capital investment in one line of activity is eschewed.

Another central feature of the Luapulan market is the strong unitary correspondence between the local structure of production and the structure of demand on the line-of-rail and other external markets. As the statistics indicate, the bulk of marketed produce is consumed outside the Province. This is particularly true of the fish trade and possibly of the commerce in other food-stuffs as well, including maize. Furthermore, locally produced commodities, including unique semi-tropical hardwoods, are exported almost exclusively in an unprocessed form.

These observations explain to some extent why the market situation in Luapula is characterized by fragmentation alongside stratification. More successful entrepreneurs rapidly attain a virtual monopoly over small market domains, yet in light of the capricious nature of policies affecting supply and demand there has been little incentive for these businesspeople to expand their scope of operations and unify the multiplicity of splintered markets. It is likely to take some time before market liberalization is taken at its face value (if indeed the loosening of restrictions persists) and entrepreneurs on all levels are willing to 'unlearn' their polystrategic behaviour in favour of greater specialization.

Behaviour and misbehaviour.

The protracted history of a politically mediated market situation has also left its impact on commercial culture. One might speak of the inherited pathologies of an imprisoned market. Heavy-handed bureaucratic controls

---

[5] For example, the 'rice boom' of the 1970s and the 'maize boom' of the 1980s.

over all aspects of daily commerce have invited corruption. The pricing distortions of the food sector metastrategy discussed above have made the smuggling of essential commodities into Zaire the most lucrative of all local businesses. In a wildly inflationary economy (also associated with the metastrategy) speculation in commodities and real estate has become far more profitable than any form of actual production.

These various 'pathologies' have not afflicted only the professional traders and persons of commerce. All strata of the population have been drawn into the distorted economic system which has defined the conditions of day-to-day survival in Luapula for many years. The following section examines the role of the various actors in the system with a view to considering how they may react to opportunities provided by the emerging situation.

The Public Sector

The Civil Service

During the latter half of the 1980s the first generation of the post-Independence Zambian civil service was nudged into retirement. Those who inherited the vacated positions were also heirs to dwindling government coffers meaning sub-standard remuneration and a debilitating lack of operational funds for virtually all government departments. Many competent professionals have opted for employment by the better-resourced parastatals or for private business careers. Increasingly, top professionals (including many of the younger beneficiaries of governmental and donor-supported staff development programs) have been poached by employers offering international standard salaries (UN agencies, bilateral donors, NGOs). Those officers remaining with government have generally found it necessary to instigate economic sideline ventures in order to survive. Often, these businesses generate more revenue than the monthly cheque from Finance.

The overall quality of the new recruits into government ministries has not been high, reflecting a rather abrupt demise in the level of teaching at the University (which has also suffered from an exodus of competent staff throughout the 1980s). Radical cut-backs in the size of the civil service are part of the SAP package, and even with the promise of better benefits for the 'survivors', the prevailing ideological atmosphere makes public service appear a dead-end for ambitious professionals.

In brief, it would appear that the line ministries do not constitute a leading force in the Provincial economy at this time, and there would appear to be little that an external agency can do to improve conditions. Still, there are many reasons why external agencies will continue to undertake selective collaboration with the support of government departments. Importantly, the *idea* of public service has not died out among Zambians, even if the hard facts of life impose more self-centred modes of behaviour. Government employment remains the main channel of social and economic mobility for those few offspring of rural families who manage to attain a professional degree. There will, of course, remain important tasks in which government action plays a decisive role for efficient commerce and social well-being; these tasks include research, monitoring and policy evaluation.

The Political Wing

With multi-party elections on the immediate horizon the prevailing single-party structures are in a very uncertain situation. The new constitution is expected abolish the post of district governor, and the status of the Central Committee within government may also undergo major changes depending on the outcome of the elections. For better or worse, the capacity of the political leadership to contribute constructively to policy formulation could be hindered for some time to come. Down on the community (ward) level, there are reports that political stagnation has contributed to a revitalization of the authority of traditional leaders (chiefs). Recognition of this could open up avenues for new approaches to the dialogue between communities and development agencies and to the development of social infrastructure where the mobilization of community participation is vital.

District Councils

Are district councils (DCs) primarily political or professional institutions? In any event, they exhibit the plights of both. Professionally, problems are those of the civil service in general (above), but to an even more serious degree. The recruitment of qualified staff is difficult, and finding truly competent staff virtually impossible. This simple fact alone goes a long way towards explaining why DC economic ventures are consistently in the red. There are basic structural problems when Councils repeatedly fail to turn a profit on the sales of the one commodity for which there is unlimited demand – bottled beer – especially considering that multi-million kwacha transport facilities have been provided gratis. By nature of the patronage system DCs have been captives of debilitating political domination. As political bodies their lot is tied to future (very uncertain) developments in the political system as a whole.

Nevertheless, by virtue of the Local Administration Act of 1980, District Councils constitute powerful corporate entities, with a formidable legislative mandate for revenue collection (although these rights are naturally worthless unless there exists taxable production or commerce within the District). Furthermore, DCs are crucial to the delivery of social services, and are likely to remain so in the foreseeable future. To sum up, District Councils are a very problematic or, if you like, challenging area for an ambitious institution-building exercise. They are not, however, something to lean on (i.e., to be given important *functions*) in a developmental investment program.

Parastatal Companies

A wide range of parastatal enterprises operates in Luapula, most of them entrusted with the provision of services on a hypothetically commercial basis. These range from UBZ, the Zimco subsidiary Indeco Small Mines (excavation of agricultural lime) and Kawambwa Tea Estate to the Dairy Board of Zambia, Lintco, ZCBC, and financial institutions such as Lima Bank. For all intents and purposes, Luapula Co-operative Union is so decisively subsidized and underwritten by the government that it also qualifies as a parastatal.

Parastatal companies are enigmatic entities. They are often relatively well-resourced (when compared with governmental departments or district councils) and operate under near-monopoly conditions. Still, in a Provincial context they are generally unreliable, seldom profitable and have a negligible developmental impact in a regional context. The problems of parastatals world-wide are well documented. These derive above all from weak management. Management problems are often connected with the patronage system and the tendency for lucrative positions to be filled with the most 'suitable' (rather than the most competent) candidates.

The regional developmental impact of Zambian parastatals is further encumbered by the low level of autonomy which provincial and district branches exercise over their local operations. The SAP program entails an agenda for parastatal reforms including a schedule of privatizing select companies. The companies to be sold off to the public will be identified in the course of 1991. At point of writing documentation is not available concerning the fate of parastatal branch offices in Luapula or elsewhere. These reforms could have important implications for the development of the economic infrastructure and the local business atmosphere, and should be monitored closely. Should branch facilities of certain companies fall under the privatization program it may behove the Finnida program to consider how these resources might best be retained (or better harnessed) in the service of regional development.

Luapula Cooperative Union (LCU)

Prior to the establishment of LCU in 1974, there were 9 co-operative marketing companies in Luapula. These had been formed soon after Zambia's independence in response to the President's promise of public support – at Chifubu in January 1965 – to anyone forming a co-operative. Four years after Chifubu (in 1969) these nine Luapula co-operatives were integrated into a FAO/SIDA 'management development' project which supplied them with management experts, transportable irrigation pumps, marketing support, transport assistance etc., for a five-year period.

The project undertook many ambitious endeavours – for example, the mass production in Mushota of irish potatoes and their marketing to the Copperbelt. These ventures folded soon after the project was terminated in 1974. The FAO/SIDA project's most permanent achievement was the establishment of the Luapula Co-operative Union as a secondary umbrella for primary co-operative bodies in the province. From its inception, LCU has consistently required external assistance to maintain its operations. First Sida then Finnida have made quite substantial investments in LCU over the years, supplemented by lesser contributions from other donors. This institution-building effort has not however had much affect on the viability of the organization as a whole. On the contrary LCU is in a chronic state of insolvency and only remains afloat by virtue of external grants. The persistent generosity of LCU's Scandinavian donors can only be understood in the light of a preconception that the co-operative movement embodies a kind of manifest destiny in the development of all rural economies. Yet contrary to the historical situation of co-operation in the Nordic world where the co-operative movement was built from the grass-roots up, the LCU 'umbrella' was established from above without a pre-existing mass movement of voluntary farmers' organisations on the primary level.

LCU's problems are not all of its own making. It has been one of many victims of the GRZ food-sector metastrategy which sought to simultaneously raise rural incomes and maintain a cheap urban food supply. As the state's agricultural budget deficit grew to unmanageable proportions GRZ defaulted on its restitution payments to co-operative Unions like LCU, which were left nursing huge debts of their own to farmers, societies, transporters, staff, and so on.

But GRZ insolvency has merely exacerbated an untenable economic state engendered by chronic mismanagement. For example, there has been no prospective investment planning to speak of; ambitious diversification programs have been undertaken for their own sake, without an eye for sustainability. Furthermore, middle management is poorly motivated when profitable branches are consistently vampirized by unhealthy loss-making operations.

It is perhaps possible that a heavy pruning exercise could salvage LCU as an organization. Were all unprofitable operations to be jettisoned, there is a chance that the remaining core of viable businesses could play a dynamic role in the province. This would nevertheless require a radical reorganization of management conditions. Just as likely, however, is that the deeply embedded culture of perverted management within the organization will prove fatal; LCU may prove incapable of instigating the required measures to eliminate dead-weight and eventually sink under the burden.

In either case, the fate of the primary societies is in limbo. Those societies which currently enjoy a degree of relative success do so by virtue of a superior marketing situation (the large Mansa societies), or because they have proved innovative in exploiting various economic alternatives: hammer-mills, independent capital formation ventures, etc. All centrally situated societies will benefit from the presumed improvements in the service infrastructure which liberalization is expected to bring about. More remotely located societies are a different matter. LCU has been assisted throughout the years on the basis of an assumption that aid will trickle down to primary producers through improved services. Neither the privatization of LCU's more businesslike functions, nor the ultimate liquidation of the organization as a whole is likely to have much benefit for remote producers. It is very possible that in such situations primary societies must assume the form of what are being called *community enterprises* within the 'new economics', small member-owned companies which actively seek out small-scale private business opportunities, rather than continuing to perform the role of agricultural marketing organizations.

The private sector

The backbone of the commercial sector in Luapula is the fish trade. Although there is little quantitative documentation on this, the general impression is that this trade is controlled by trader/transporters whose main economic interests are on the Zambian Copperbelt. This is a classical case of the simple extraction of primary produce: local processing is minimal and the bulk of the profits are realized (and accumulated) at the retail end of the operation on the line-of-rail. The developmental impact is heavily limited by the mode of extraction, as well as by the fact that the artisanal fishery lacks repositories for the saving and investment of profits.

Resident large-scale private businesspeople in Luapula are few. Aside from the 14-15 individuals considered to be farming on a 'commercial' level, the few prosperous business folk are concentrated in the service (retail) sector (Spider Machisa, Patel Syndicate, etc.). Private manufacturing and processing 'industries' are virtually non-existent, with one important exception: traditional breweries.

As business enterprises, traditional breweries are fairly representative of the mass of small-scale or 'informal' ventures in Luapula. Breweries are family operations, generally managed alone by an adult woman. They coexist with other economic activities as part of a polystrategic mode of livelihood. The production process is technologically simple, yet it embodies certain specialist knowledge which is not universally available. Turnover is rapid and profits are utilized immediately, partly as investment in a new batch and partly as general consumption. There is no capital accumulation. Limited demand is one constraint on the expansion of production, but the lack of raw materials (millet, maize) or capital (to purchase additives) constitutes more substantial bottlenecks. Production is not always strictly market-oriented: beer can be traded for labour, or used to meet past debts or create future social obligations among relatives and neighbours. Bookkeeping is uncommon, profitability is seldom quantified, and there is little if any dependence on (or access to) external sources of capital.

There are many reasons for the underdeveloped nature of the Luapula business community. Governmental sources see weak management as a major constraint. Aspiring entrepreneurs themselves cite the lack of capital, price controls, high operational costs, variable demand and the unreliable supply of commodities as the prime obstacles to launching a business. Obviously there is no one obstacle to the development of private business.

The situation is complex, and probably appears more complicated than it is because of the near total lack of reliable research data.[6] Nevertheless, two factors may be singled out as causally prime: the unreliability and high cost of supplies (parts, raw materials, retail goods), and the lack of reasonably priced working capital.

Lack of capital for initial investments in capital goods is in principle an artificial problem created by the lending policies of local financial institutions. Local branches of the large commercial banks (Barclays, ZNCB) apply the same set of assessment criteria as do their headquarters in Lusaka. Most aspiring small-scale entrepreneurs fail to qualify for formal credit: they lack training and business experience, and above all they lack collateral.

The problem of unreliable supplies for commodities is more fundamental, perhaps, in the sense that in many cases the desired commodities are simply not available on the Zambian market at economic prices. But often the decisive bottleneck is not at the production nor import stages, but is one of affordable acquisition and timely transport to or within Luapula. It should furthermore be noted that the two main causal factors identified above are interconnected: an inflexible capital market inhibits the establishment and development of transport ventures which in turn contributes to bottlenecks in the supply of commodities.

The Middle Ground
Zambia's economic problems have undermined the standard of living in rural communities and the on-going transition cannot be expected to provide much short-term relief. On the contrary it is not realistic to assume that the measures needed to eliminate the chronic budget deficit will the lead to the improved delivery of basic public services. It would benefit the welfare of rural people therefore if local communities could strengthen their organizational infrastructure and with it their means to articulate objectives and initiate organized action to realize these goals. Donors and NGOs alike have expressed the wish to facilitate the consolidation of such organizations. However, they have been hard to find. In many cases they have had to be created on the initiative of a funding agency. The question arises: can this be done without compromising the basic criteria of autonomy? Here again the contradictory yearning for liberalization vs. intervention presents us with a paradox.

Luapula is not completely replete of local organizations, however. Be-

---

[6] The only survey of the Luapula business community available was conducted in 1974 (Baylies 1978).

tween public sector institutions and private business there lies a grey territory of economic activity which combines elements of both. While neither particularly large nor significant in terms of its economic functions, 'third column' entities of this sphere attract interest precisely because of their hybrid, ostensibly 'transitional' nature.

Within this sector fall community-level economic organizations, above all those primary co-operative societies which have achieved a level of administrative autonomy vis-à-vis LCU and the Department of Marketing and Co-operatives. Here are also found those few non-governmental organizations which can be found in rural Zambia, the most important of these being religious congregations. Certain donor-instigated activities might also be catalogued here; for instance, some aid projects are so loosely integrated into any local administrative structures that they can hardly be considered governmental, yet their mode of operation is clearly not commercial either.

The economic plight of these local organizations is that the possibilities of developing their resource base are structurally constrained. Their inconclusive legal status places them outside established channels of access to private sector resources. Inasmuch as they submit to the conditions of public or donors' support they forfeit their delicate autonomy. In any event, such local organizations have not proved conducive to the running of business activities by virtue of their social makeup.

Community-based institutions naturally mirror the social and political relations of their environments. Rural communities are not 'democracies' in the Western understanding of the concept. Structures of authority appear very conservative, androcentric and elitist. Yet the potential strength of such organizations, that being their capacity to provide an outlet for community sentiments, resides in their autonomy, and this very autonomy provides a social context – a site for political struggle – within which the constitution of local authority can be redefined.

As this brief overview suggests the main character of community-based corporate bodies is political. Their economic role is subordinate. The political processes which these organizations embody, however, are crucial for negotiating the role and status of interest groups within the community. The development of the private sector will stimulate the emergence and consolidation of some social groups, as well as the demise of others. Drawn-out and indecisive struggles between nascent and moribund interest groups can cripple a community's ability to cope with changes in the national economic and political environment. In terms of the policy of intervention in local communities it would appear advisable to concentrate any efforts on improving the economic conditions for competent entrepreneurs to carry out valuable economic functions in their communities while maintaining a healthy distance from the political fora within which power is under negotiation and the political culture is being recreated.

A Strategy for Economic Transition

The process of transition from a centrally controlled economy to one in which market forces have greater play seems to be underway in Zambia. For the moment, the speed and indeed the permanence of this tendency in Luapula remain shrouded in uncertainty. Still, the trend is clear from the three perspectives of ideological analysis, policy statements and empirical observation. The province's major development effort, the Finnida-sponsored LRDP, should consider adapting its operations to these realities in a number of ways: in the structure and content of its investment program, in the selection of the actors and institutions with which it collaborates (and the forms of organization within which collaboration occurs), in the mode of its situational analyses, and in the style and fora of its policy discourse.

Perhaps the most sensitive issues concern the continuing relationships among the public sector institutions of the Zambian government, provincial administration and the co-operative movement. Most important here, perhaps, would be to articulate a provisional conception of the specific aspects and dimensions of market liberalization and private sector development that the Finnida investment program hopes to promote. The formulation of a development strategy in these terms would naturally require some explicit thinking about the relationship between economic goals and social costs, and about Finnida's commitment to considerations of equity as a complement to developmental issues. Such a conception could then provide a basis for reassessing the program's intervention modes and targets, as well as for defining the necessary scope of its involvement in the public sector in the future.

The discussion above has shown that the task before rural Zambia is not so much of 'freeing' the market, but of consolidating and strengthening it. Necessary (if not in themselves sufficient) prerequisites for this are strong financial institutions and a greater local integration of supply and demand. This latter point has several dimensions:

- in the more immediate future, the development of retail distribution networks for both consumer and capital goods is of primary importance;
- in a broader time-frame, the linking up of primary production with secondary manufactures/processing is the primary developmental goal. This naturally has implications for *inter alia* the structure of primary production (e.g., 'improving local agricultural systems'), for marketing, and for the financing of the secondary sector.

These strategic elements imply the following points of entry:

Primary Production

Priorities within the sphere of agricultural development relate to the incremental enhancement of food production for consumption and the local market, including nearby urban centres. The nature of the obstacles to progress in this sector is on the one hand biological (varieties) and cognitive (techniques), and on the other hand tied up with the problems of the general marketing infrastructure. The biological and technical issues require long-term systematic research. Research geared to the improvement of genotypes and production methods has already been going on for generations on the farmers' own fields, and this indigenous knowledge should provide the basis for further inquiry. The Adaptive Research Planning Team (ARPT) has been pursuing this approach but has suffered from administrative marginalization. Work of this kind is extremely time-consuming and labour intensive and the *economic* benefits if assessed in terms of marketed surplus are not significant. An ARPT-type program only makes sense if it contributes to a revitalization of the overall thrust of mainstream agricultural research. As an administratively isolated parallel effort it is in danger of becoming a cul-de-sac. The marketing of agricultural produce is not a special case but will be discussed in connection with general problems of distribution below.

Fisheries development appears to be too large and complex an area for any single agency to handle. GRZ has failed to elaborate a systematic program which would simultaneously tackle the bio-ecological, social and economic dimensions of artisanal fisheries development. Impetus for interventions on the scale necessary to make a qualitative impact is low since the fish does, after all, seem to be reaching the market. While limnologists have signalled for decades that biological production is in a slow but immutable state of decline, the only urgent questions which seem to arise from the casual observations of donor missions are of a largely social nature (concerning health and mobility). There is no question that there is untapped potential in the lakes and swamps and that members of the fishing communities who supply so much valuable animal protein to urban residents are inadequately compensated. It is nonetheless uncertain whether a developmental intervention into the fisheries extensive enough to rectify the situation is at all conceivable. Assistance of a humanitarian nature to improve the delivery of social services, especially to the Bangweulu swamplands is morally justified.

Secondary Sector

The main structural weakness of the regional economy in Luapula is the underdevelopment nature of its secondary (processing, manufacturing) sector. As noted above, primary produce is generally exported in unprocessed form. This is a fertile area for the development of private business ventures under the new liberalized policy atmosphere. The key obstacles to the development of secondary industries in the province have been, it is argued above, weak capital markets and unreliable distribution networks. These are discussed below. In addition there are sundry subsidiary problems which could be addressed through development interventions if suitable channels and modes of operation can be identified. These include the development of business culture through management training, and the catalyzation of technological progress appropriate to local economic and cultural conditions. Luapula has a basic infrastructure of public training institutions through which such technical goals could be pursued. Since these institutions have long suffered from the general dilemma of the public sector discussed above, significant physical and organisational renovations would be required before they could deliver the desired services efficiently. While this might be the most viable alternative in the long term it is worth considering other options as well.

Urban-based enterprise can reasonable be expected to constitute the leading sector in the development of secondary industry. However, the dynamism of the regional market will depend to a great extent on the 'decentralization' of technical support services into more remote rural population centres. No rural distribution network can survive unless certain general mechanical maintenance services are also available outside the main administrative centres. It is reasonable to postulate that adequate technical expertise exists in the rural communities; it is uncertain whether viable businesses can emerge in these settings without preferential support. Lessons of the past decades are in any event unanimous: such services can only be delivered through private initiative. Artificial corporate entities (for example, primary co-operative societies) cannot manage to maintain such ventures. Under some circumstances, rural church congregations might have the required organization and discipline. What these conditions are cannot be said precisely since this issue is poorly documented for the time being.

Distribution

It is obvious that the strategic role of middleman functions will become ever more crucial under liberalized market conditions. Trader-transporters are becoming key players in the development process. The strong Finnida presence in the transport sector means that future investment decisions can have a significant impact on the way in which the sector develops. A Finnida move to withdraw from the sector, or to privatise its existing assets, would have a radical impact on the structure of distribution networks as would an opposite decision to continue support to the LCU transport 'monopoly'. The main argument for continued support to LCU can be made on equity grounds. Left to free market forces to determine the structure of distribution networks, many remote communities would be completely cut off from reliable transport. While this is an important consideration, the arguments against further support to LCU via its transport wing are also strong. First, LCU represents precisely the sort of organizational inefficiency which has been diagnosed as the main problem of the Zambian economy. It is hard to rationalize the continuation of subsidized investments in LCU. Second, investment in LCU can hinder the profitability of private transport. This is counter-productive in relationship to the overall strategy. Third, the future development of the transport sector must reflect market developments in the region as a whole. LCU's commitment to a specific market and a certain clientele is also an encumbrance.

The question then is of how to avoid the worst negative consequences of liberalization in equity terms while stimulating private initiative in the development of the distribution system.

Financial Institutions

Donor agencies in the province (Finnida, Sida, GTZ, IFAD, Unicef) have all noted the need to make cheap capital available to prospective entrepreneurs. This aim has resulted in the introduction of special credit arrangements, usually in the form of revolving funds under the administration of a specific project or the co-operative movement. Perhaps some funds have also been channelled through the parastatal Lima Bank (ex-Agricultural Finance Company, etc.). These credit lines have generally entailed special, non-commercial arrangements concerning lending criteria (no collateral required), (below market) interest, (subsidized) administrative costs and repayment (often in kind). Very few of these special borrowers have graduated to the commercial capital market which continues to exhibit extremely inflexible lending policies.

The liberalization of trade requires the liberalization of capital markets as well. This can perhaps best be achieved through agreements between the donor agency and pertinent (private) financial institutions. Such agreements would be intended primarily for entrepreneurs in the manufacturing, processing and transport sectors of the local economy, not for seasonal agricultural credit. Discussions should also be had concerning develop saving and investment possibilities and joint ventures involving donor, bank and entrepreneurs if necessary.

Actors, Institutions & Forms of Organization

The focus in the emerging economic situation is on private entrepreneurs. This is quite distant from the policy underlying the program as it is now running. Finnida is quite committed to collaboration with organizations and institutions which have become quite dependent upon a Finnida presence. An abrupt phasing out of investments in these public sector institutions – LCU and the Ministry of Agriculture – would not be an act of good faith, nor would it instil confidence among other actors in Finnida's commitment to long-term collaboration.

A possible solution might be to diversify the investment program by seeking new forms of collaboration in the form of joint ventures between various Zambian actors – such as local investors, financial institutions, the Provincial administration, LCU – and Finnida or other external agencies. There are no doubt major legal problems involved in such ventures; the political obstacles may also be formidable. In the final analysis it boils down to the question of how active a role the donor wants to play in the restructuring process. Only innovation brings about change. If Finnida as the major investor is content to 'adjust' to a changing environment rather than take an active role in seeking working alternatives it cannot be faulted. On the other hand, why fuel a ship with no destination?

4

MARKET LIBERALIZATION &
SMALL-HOLDER LIVELIHOODS
Luapula households before 1988 and after
the 1996 economic reforms
(1997)

Introduction

Under the Third Republic, which commenced in late 1991, Zambia has become a flagship of IMF-style economic liberalization policies. Since taking power, the MMD government has carried out an aggressive program of economic reforms. These reforms have sought to reduce the volume of the bloated public sector economy it inherited from the UNIP era. The reform program – which has been hailed as 'bold' and 'radical' by proponents, and 'harsh' and 'destructive' by its detractors – has entailed classical liberalization measures: the privatization of public and parastatal enterprises, civil servant retrenchment and the dismantling of agricultural subsidies.

The shifting policy environment

During the era of UNIP's Second Republic (1973-1991), Zambia (and its international backers) invested heavily in stimulating agricultural output. A central mechanism to this end was the promotion of hybrid maize via a package of input and marketing subsidies. This program was relatively successful – dramatically so in parts of the country (the Northern, Luapula and Northwestern Provinces) where hybrid maize technologies were not already common. Small-holder maize production in these regions rocketed between 1975 and 1989. Alongside input (seed, fertilizer) subsidies, govern-

ment policy channelled massive support to grain marketing via co-operative organisations. All across the country, and especially in the new centres
of the 'maize boom,' farmers were encouraged to establish co-operative
marketing institutions at the primary (community), district and provincial
levels. Co-operative societies became the prime conduit for the channelling
of subsidized credit, input supply and marketing services to a new breed of
'emergent' rural small-holders.

The UNIP government also subsidized the cost of milled maize flour
('mealie-meal') to urban consumers. Taken together, these expenditures
burdened a national economy that had been living beyond its means since
the demise of copper prices on the world market in the 1970s undermined
the basis of Zambia's public revenues. By the mid-1980s Zambia had the
world's largest per capita foreign debt, and her credit standing with international financial institutions was rapidly evaporating.

Zambia's history with the 'stabilization' and 'structural adjustment'
family of economic reforms dates to the mid-1970s when the Zambian
government began its collaboration with the Bretton Woods institutions.
Zambia's first financial assistance under the auspices of a 'structural adjustment program' began in 1983. This program entailed a massive devaluation
of the Zambian kwacha with the introduction of a currency auction in 1975.
The government was also called upon to freeze wages and cut the budget
deficit. The system of subsidies surrounding maize was a constant bone of
contention. A reduction in maize subsidies was an obvious requirement to
lowering the budget deficit, but the UNIP government's attempt to cut back
consumer subsidies for mealie-meal in 1986 was met with urban street riots.

In 1987, Zambia broke with the IMF and launched a home-baked New
Economic Recovery Program. The kwacha was revaluated, but economic
performance continued to falter. By the end of 1988, rampant inflation and
a growing undercurrent of political unrest drove the UNIP government
back to the IMF. The UNIP government's Policy Framework Paper (PFP) of
1990 avowed a renewed commitment to market liberalization. The new
policies, which included the decontrolling of the grain market and the
reduction of input and transport subsidies in the agricultural sector, were
just taking effect when MMD ousted UNIP from power in the 1991 elections
(Mutukwa & Saasa 1995).

MMD based its election campaign on promises of 'liberalization,' which
was widely understood to mean two things: the legalization of private

commerce and the end of UNIP's political monopoly. In the context of the rapid deterioration of the government-controlled economy and the growing alienation of the political leadership from popular demands, both of these promises enjoyed great popular support. After its resounding election victory in 1991, MMD reaffirmed its commitment to market reforms, and accelerated the cautious pace initiated by UNIP.

Under the co-operative/parastatal grain marketing system that prevailed through the 1991/92 season, the bulk of the maize produced by Zambian farmers was marketed through local co-operative societies. Within the new policy framework which emerged after Zambia returned to the IMF fold, private traders were to have greater access to rural markets. Private participation in agricultural marketing was constrained, however, by the predominant role that the dense network of co-operative institutions had in credit supply and marketing services. The abrupt dismantling of agricultural subsidies and the decontrolling of grain prices under MMD caused the co-ops to lose their competitive edge over private traders. The drought of 1992/93 left little maize to market, but already in the 1993/94 season about half of small-holder farmers sold their maize to private dealers whereas only about one in ten marketed through a co-operative society (Chiwele et al 1997: 19).

Analytical framework
This study looks at the consequences of the economic reforms implemented by the MMD government for the rural poor. Given the economic and political importance of the agrarian sector in Zambia's future, it is surprising that the analysis of the impact of the agrarian policy measures on small-holder rural households rests on such a thin empirical basis. Few studies undertake to analyze the consequences of the economic reforms. This is partially a reflection of the poverty of Zambia's independent research sector. Contemporary research is strongly dictated by external interests. For the moment, expatriate academic studies, mirroring the concerns of the major donors, are far more interested in 'governance' than in the state of the rural economy.

Those studies that do address the impact of market liberalization are caught up in a highly politicized field. One set of presentations basically rejects the political legitimacy of the economic reforms (or their particular mode of implementation). These are grounded in a political agenda that has become increasingly critical of MMD rule, and are generally carried out as short-term consultancies in the 'non-governmental' sector where they circulate as 'gray' literature.[1] These critical studies focus on the 'social' dimension of the reforms, highlighting distribution problems of social polarization and gender inequities. Two central claims are generally put forward in this literature: first, that economic reforms have seriously undermined rural welfare, and second, that as a group, women have been hit the hardest by the negative impact of market liberalization.[2] Methodologically, these studies are largely based on the 'quick and dirty' survey techniques of the development consultant, and their conclusions are vulnerable to claims of ideological tendentiousness. Pressed to offer 'recommendations,' the authors of these studies are obliged to generalize on the basis of non-contiguous data, or to assess the impact of liberalization through vague reference to a pre-existing 'un-liberalized' situation which is not empirically defined.

Another set of studies, largely carried out under ministerial mandate as a 'monitoring' exercise, accepts the logic of the reforms as unproblematic. The terms of reference for such studies are built upon the (neo-liberal) assumption that rural producers are major beneficiaries of market liberalization, and that free markets will provide the needed incentives to improve the efficiency and yields that the parastatal marketing system suppressed. Work in this vein is predominantly economic (as against social) in focus, and expresses an instrumental concern with prices and margins, market information, and private sector response to liberalization. Such studies are generally professionally competent, but spatially myopic. For example the impact studies prepared under the auspices of the Department of Agriculture tend to look at situations, such as the Eastern and Central Provinces, or Ndola Rural, where 'the market' is relatively well articulated and where there is a long history of commercial activity. The more marginal regions (again, Northwestern, Northern and Luapula Provinces) where national or regional markets in agricultural produce were virtually non-existent before the 'maize boom' do not figure in these studies.

Ideally, an overall assessment of the impact of economic reforms would synthesize pre- and post-liberalization data drawn from a variety of situations. Lacking the data (and other means) to carry out such a broad synthe-

---

[1] Representative of this literature are the provincial assessments prepared by the Danish organization MS.

[2] For example, L. Haddad et al (1995: 893) argue that economic adjustment policies put women 'at a distinct disadvantage' compared to men. The authors qualify this claim, however, with the observation that 'many women are simply spectators rather than innocent victims of economic reforms.'

sis, this study compares the production and marketing behaviour of small-holder farmers in a single rural community in the 'marginal' Luapula Province before and after the economic reforms of the 1990s. Based on an identical data set compiled in the same localities in 1988 and again in 1996, the study seeks insights into the following questions:

- What distinguishes the post-reform market from the state monopsony?
- Does market liberalization provide incentives for improved productivity?
- Who benefit from market incentives?
- What factors affect the selectivity of opportunities in the liberalized market?
- What has been the gender and inter-household impact of market liberalization?

An important theoretical issue concerns identifying the direct consequences of market liberalization.[3] The domestic economies of rural Zambia have been in a state of flux for the entire century, and it is difficult to discern which of the observable changes between 1988 and 1996 can, in fact, be traced to the economic reforms. A related conceptual problem arises from the fact that the impact of 'liberalization' has been mediated by a number of intervening factors. Unlike a number of countries (and Zambia herself in the 1980s), economic reforms have actually been put into effect rather effectively in the Third Republic. Still, donor agencies and local politicians have each, in their own way, intervened in the relationship between the producer and the market, affecting the expectations and behaviour of farmers and traders alike. In some cases, the relationship between the reforms and a behavioural shift will be relatively clear and self-evident; in other instances, it will be best to merely note that a given shift in behaviour occurred coterminously with market liberalization.

A comparison of the pre- and post-reform situation should offer an opportunity to test empirically the relative effectiveness of a state-controlled vs. a free market situation. In certain respects, it does. In 1988, the maize market (but not the market in groundnuts, for example), was effectively regulated by the Zambian state via the parastatal institutions of the co-operative movement. By 1996, state (or parastatal) interference in rural Luapulan markets was virtually non-existent. One must still keep in mind that the liberal situation of 1996 is in no sense in a 'state of nature.' It is not

---

[3] See e.g., D. Booth et al (1993) for a discussion of the distinction between liberalization in policy and practice.

so much a 'free' as a 'freed' market, and must be approached via an under-standing of residual path dependencies. Any analysis of the contemporary situation must bear in mind that the farming systems, technological biases, consumption patterns and social relations of the liberalized nineties all bear the imprint of the 'illiberal' eighties (and indeed of much earlier times). This limited exercise will hopefully refine the questions we ask about the boons or burdens of market reform, but it should not be taken as a test-case of the market-driven development model as a whole.

Research Site[4]

The fieldwork on which this study relies was carried out in Chief Milambo's area, Mansa District, Luapula Province, Zambia (Figure 1). Milambo is locat-ed about 80km (or 90 minutes to three hours by gravel road) from the provincial boma of Mansa. In ideal-typical terms, there are two modes of being rural in modern Zambia. A locality is either integrated into the broader regional/national market context, or is economically marginal. In practice there are many gradations of market integration and marginality, and one of the outcomes of this study of Milambo will be to examine how economic reforms have affected and redefined the shifting boundaries between these two notional modes.

For the purposes of this analysis, Milambo is taken as a 'marginal' locali-ty. It has not always been so. From early decades of this century until the mid-1970s the Milambo economy was a 'dependent' appendage of the line-of-rail copper-based production enclave. Dependency was expressed in the spatial dislocation of production and reproduction. That is, the wealth upon which the standard of living in Milambo depended was generated by men (and women) laboring hundreds of miles away from the land in which Milambo society is rooted.

Land embodies the cultural and spiritual foundations of Aushi (or Ushi) society; the spirits of Aushi forefathers reside in the land, and the clans and lineages which constitute the backbone of Aushi social organization are tied to a specific spatial configuration of usufructory land rights. Land use decisions are largely managed by men (who occupy the vast majority of the village headmanships responsible for land allocation). Yet the cultural system governing the control of resources is built upon the principle of 'mother-right.' In Milambo, like much of central Africa, clan and lineage membership – and hence, descent and inheritance – are reckoned matrilin-

---

[4] For more details about environment and demographics of the Milambo area, see Annex 1 to this chapter.

eally. A man must commonly approach his wife's relatives (on her maternal side) for his initial allocation of land after marriage. Without overstating the case for women's entitlements, it is evident that women have greater means to influence the use of material resources in matrilineal Luapula than the prevailing literature on disempowered African women suggests. The pattern of labour migration to urban areas, instigated by colonial tax laws, split households in two, as men left their wives' villages in search of wage employment. Still, in matrilineal Aushi society the foundations of the matrifocal family remained intact even when the husband and father was absent. Dependency did not sever the people's roots in the land, but spatial dislocation did mean stagnation for Milambo because the vital unity of land and labour was shattered.

Over the past 50 years, the Zambian state launched a series of meagre attempts at stimulating local production during the extended period of dependency. These failed, both in the late colonial period and in the early years of the independent Zambian government. These interventions were half-hearted, and thus unsuccessful, because neither the colonial nor the post-colonial regime was particularly interested in creating an economic environment which would have made local agriculture more attractive than migrant labour.

Already during the 1950s, however, urban politics and technological developments in the copper industry began to undermine the demand for cyclical migrant labour. Throughout the sixties, even before the price crashes of 1971 and 1975, the intake of short-term contract workers was on the decline. Dependent labour reserve areas like Milambo were gradually becoming marginalized with respect to the needs of the line-of-rail. President Kaunda's appeal at Chifubu in January, 1965 to 'go back to the land' expressed the situation succinctly: migrant workers were no longer needed as a buffer against labour shortages; their labour was now needed back in the villages producing food for a stabilized urban proletariat (Quick 1975: 153).

The recession which set upon the copper industry following the price crashes of the 1970s consummated the marginalization process in Milambo. Milambo became a marginal economy in two senses: on the one hand, there was only a marginal demand for local commodities (including labour) outside the immediate community. The corollary of this is that surplus production beyond local demand played only a marginal role in producers' strategies. The maize boom of the eighties promised to transcend this marginalization; Milambo appeared to be undergoing a kind of re-integration into the national and regional economy, now via the marketing of local agricultural surplus. The qualitative significance of this process lay in the fact that for the first time since the colonial annexation, the spheres of production and reproduction of the local economy were experiencing a reunification, albeit incomplete and one in which dependence on the line-of-rail persisted through various forms of subsidy and political control.

Milambo is not a 'typical' marginal locality. It is the site of the Senior Chieftaincy of the Aushi people, one of several matrilineal Central African 'tribes' inhabiting Zambia and Zaire. Under the colonial regime of Indirect Rule, Milambo's chieftaincy was the seat of the Ushi-Kabende Native Authority. Two large Christian missionary stations (one of the Roman Catholic White Fathers, and another of the Protestant Christian Missions to Many Lands (CMML)) were established in the chiefdom in the 1930s. During the sixties, the independent Zambian government instigated what was then the largest co-operative farming scheme in Luapula in the northern reaches of Milambo's area. Yet for all these links to the larger world, Milambo's Aushi have lived their lives largely in the margins of the national, Copperbelt and line-or-rail economies. In terms of certain basic indices – above all those of physical, commercial, and local organization infrastructure – Milambo exhibits the classical hallmarks of marginality, and in this sense, the rural inhabitants of this area are representative of a large section of Zambia's rural population.

An examination of Milambo reminds us, however, that marginality can be a relative and dynamic state. During the fluctuating copper booms of this century, Milambo was a heartland in terms of labour recruitment, and in the 1980s maize was not a marginal crop. One intention of this analysis is to examine the extent to which economic reforms have enabled or constrained nascent reintegration, and whether market liberalization has contributed to 'development,' either via the consolidation of agrarian capitalism, or in any other sense.

Methodology[5]

The comparison of the pre- and post-liberalization situations is based on a carefully sampled set of 200+ questionnaires collected in 1988 at the height of the maize boom, and in 1996, in the fourth season of liberalized grain marketing in Zambia. The 1988 interviews were carried out under assuranc-

---

[5] For more details about the sample and study design, see Annex 2 of this chapter.

es of anonymity, and it was therefore not possible to re-identify the sampled households for the 1996 restudy. Hence, it was not possible to examine the impact of market reforms on individual households. Still, the 1996 sample was constructed on the basis of the same parameters, and in the same villages, as the 1988 questionnaire, and if one is willing to assume that the behavioural patterns under study have a structural or collective logic, the outcomes of the two studies should be comparable, if only in a heuristic sense.

One can and should regard quantitative questionnaire data with healthy scepticism. There are many factors which interact to weaken the quality of information obtained through standardized questionnaires in any context, and the conditions of working in a remote African rural locality amplify these. A great deal of care was taken to ensure that the questionnaire was comprehensible to respondents and easy to administer in a reasonable time. The interviewers were professional enumerators with many years of experience of working in similar localities and conditions. Sampling and interviewing techniques were polished prior to the field exercise. Still mistakes were made, especially in the 1988 cycle, and some of the data had to be discarded. My own predisposition as an anthropologist is to favour qualitative and experiential data, and I would not feel comfortable using and interpreting these materials did I not have the benefit of many months of residence in and recurrent visits to Milambo as a basis for this analysis. In the course of the ensuing discussion, I try to signal the issues and areas in which the quantitative materials can, at best, provide a basis for speculative and inductive reasoning. At the end of the day, however, I feel that an exercise of this kind is justifiable in both practical and scientific terms. Without such a systematic survey, it is not possible to gain an overview of the basic economic structures at work. This is truer than ever in the current situation, where the government has more or less abandoned any attempt to maintain valid statistics on economic indicators in the liberalized rural economy.

The Domestic Economy

This section explores the changes that have occurred in the way that people in Milambo secure the means of their everyday existence within the domestic sphere. As in most agrarian communities, the domestic economy in

Milambo revolves around agricultural production. The farming systems in Milambo are complex, and in a continual state of evolution as producers respond to changes in the economic and natural environments. A hundred years ago, Milambo farmers based their production on citemene (swidden) cultivation, which kept production units based on the extended family in a mode of semi-nomadic residence. Villages were established for only a few years until the forest cover in the immediate vicinity was depleted by swiddening and people moved on in search of full-grown trees. The main crop at that time was millet, which was gradually giving way to the current staple of cassava. The advent of colonialism (from the end of the nineteenth century), and the system of cyclical labour migration to the copper mines in search of waged income drained Luapulan villages of male labour. This contributed to the consolidation of cassava (a crop with a relatively low labour demand) in the farming system. The colonial system also enforced more sedentary settlement, and the development of roads and a social service infrastructure drew people into roadside villages (a mission hospital and school began operation at Lwela in the 1930s; a post office, local court and police station joined these in the 1950s). A fall in child mortality and a general reduction in the mobility of the population increased pressure on the land and pushed producers to find more efficient means of crop production.

The external shock of the crash in copper prices in the mid-seventies also had important implications for the domestic economy. The drastic contraction of the mining sector shoved unskilled migrant labour out of the urban wage market. This had dramatic consequences for Milambo. Many of the rural households that had come to rely on cash remittances from migrant members over several generations were now faced with a serious reduction in cash incomes. What is more, it grew apparent to young rural men that the urban labour market was an increasingly tenuous option. By the end of the 1970s, a significant number of local farmers were looking to agriculture for a means to compensate for the cash squeeze arising from a growing population and reduced cash income opportunities.

The arrival of chemical fertilizers and hybrid maize seed in this area in the 1980s offered an attractive option for 'progressive-minded' farmers to improve their cash incomes. But even before the advent of chemo-genetic farming technologies, local producers had devised a variety of ways to ensure food security. Thus, one finds a multiplicity of crops and cropping systems within the area. In addition to millet and cassava, a typical household will produce groundnuts, beans, pumpkins, sweet potatoes, a range of leafy greens, sugar cane, maize, tobacco and other crops for both personal consumption and exchange. Cultivation takes place on a number of different categories of field type. These might be thought of in terms of their relative age, ranging from the new-swiddened (umunda) plot, to semi-permanent (ibala) fields. People also exploit the water-logged dambo areas in various ways, as (ilungu) gardens for vegetables or, more recently, for rice production.

The local market for primary produce is limited, but important for households with specialized needs or a limited capacity for their own production. Much of the local trade occurs in the context of reciprocal and barter trade within and between households. The majority of the local production is either consumed or exchanged in its primary state. Local

Source of Income	1987 %	1995 %	Change
Sales of crops	89	83	-6
Sales of beer	55	50	-5
piecework	26	17	-9
Sales of fish	21	27	+6
trading	14	12	-2
Sales of crafts	13	8	-5
Sales of other produce	11	11	nil
Salaried employment	10	5	-5
Sales of livestock	7	2	-5

Figure 1. Proportion of sample households reporting income from common economic activities. Source: survey data. Note that the above order does not necessary represent the relative proportion of each activity in the aggregate 'GNP' of the Milambo area; it only expresses the number of households involved. For example, few households may be involved in trading, but their combined incomes from this commerce may well surpass the combined annual turnover of the fishing 'industry.'

industries for the refinement of primary produce are few, beer brewing and artisanal trades are the major exceptions. The following section examines the structure of household livelihoods and the changes that have taken place in connection (or coterminous) with the liberation of the grain market.

Livelihoods

Figure 1 indicates how many respondents to the survey obtained income from the most common economic activities in the study area.

Figure 1 reveals some basic facts concerning the structure of the local Milambo economy. Commerce in agricultural produce (maize, millet, groundnuts and cassava) appears to command the economy. The processing sector is dominated by the brewing of millet beer, the community's only agro-industry. The Chiefdom's economic marginality is underlined by the low activity level in general commodity trading. By way of contrast, Allen found trading so important in Mabumba's chiefdom, just north of Milambo but intersected by a major tarmac road 30 km from Mansa, that '[o]ne could almost call Mabumba a retailing economy for which farming is mainly subsistence.' (Allen et al 1988: 50) The situation in Milambo is clearly quite different.

Within the context of a basically agricultural system certain minor specializations appear to thrive. The most important is fishing. A second specialization involves traditional crafts. These comprise such diverse activities as carpentry (doors, door and window frames, coffins, and furniture), masonry, basketry (including grain storage bins), musicianship and herbalism (including witch-finding). Wage employment is limited to a thin upper stratum of the population, many of whom are civil servants. The government also offers a limited amount of temporary wage employment, largely road maintenance. Competing with the government for largest wage employer is the Lwela CMML mission which also pays the salaries of many of the hospital staff.

Changes in Livelihood Patterns

The economic reforms have coincided with a general shrinkage of the local market: most dramatically with regard to crop sales, salaried employment and casual labour. Of these, the reduced number of households benefiting from crop sales and of opportunities for casual labour would seem to be directly linked to liberalization. The decline in wage employment probably has less to do with civil servant retrenchment than with the departure of a resident expatriate CMML missionary in 1989. When the station was manned by expatriate missionaries, CMML was the largest local employer in the community. The veritable disappearance of a market in livestock is a consequence of a chiefly ban on goats and pigs in response to complaints by cassava growers over damage to their fields. Plausibly, reduced opportunities for crop and casual labour incomes have led to increased fishing and fish marketing. Given the limited amounts of fish available in the local waters, the fish market is quite localized and on a very small scale.

Economic Strategies

The transition from state monopsony to liberalized markets is only one in a series of external shocks to which Milambo producers have been exposed over the course of this century. The first, and in many ways the most profound of these shocks was the colonial annexation of north-eastern Zambia after the turn of the century. The colonial administration imposed taxes on the local population which could only be paid in cash, thus necessitating labour migration. A second major impetus to change came with independence from Great Britain in 1964. Independence brought with it the post-colonial African party-state which sought to ground its authority in networks of clientelist relations with local supporters. In its early years, the post-colonial Zambian state under UNIP cultivated the ideology that development could only emanate from the 'Party and its Government.' In popular parlance, clientelism has been equated with 'spoon-feeding,' and a culture of passivity which undermined local initiative. The elaborate package of agricultural subsidies associated with the Lima Program has been seen as one manifestation of UNIP's efforts to cement the clientelist dependency between rural citizens and the state.

The post-independence period has seen two major crises affecting lives in Milambo directly. The first was the sudden contraction of the mining sector in the 1970s, which cut off the labour reserves like Luapula urban wage remittances and focused economic expectations on agriculture. The withdrawal of agricultural subsidies as discussed here is the second such shock. Coupled with the vagaries of health, politics and climate, these shocks have taught rural producers the wisdom of flexibility in their economic strategies. Talking with farmers and observing their actions over the course of many seasons, it is evident that actors are hesitant to concentrate their resources (land, labour, and capital) in any given enterprise, be this a cash crop or a commercial venture. A rule of thumb is to try and hold some option open in case of economic failure or disappointment, and to keep some part of one's assets mobile in case an unforeseen opportunity appears

	Proportion (%) of respondents	
Sources of Income	1987	1995
0	2	5
1	18	22
2	37	40
3	26	23
4	13	8
5+	4	3
	n=221, mean=2.42	n=214, mean=2.15

Figure 2. Number of different income sources reported by respondent households.

on the horizon. While flexibility may be a wise strategy, it is also a delimiting one. A lack of specialization and the tendency to spread one's resources over a range of different activities constrains investments in technological development and is thus an obstacle in the path of improving land and labour productivity.

For the majority, even strategic flexibility is an unachievable ideal. Agricultural self-employment, supplemented by various sidelines in trading or beer production, is by far the most prevalent profile of a Milambo producer. This can be observed in Figure 2, which represents a crude index of the complexity of economic strategies.

The majority of households combine several activities in their pursuit of cash income. Strategies manifest choices between social and economic options, and these choices are constrained, among other things, by an individual's (or household's) resource endowment (Ellis 1988). The most cogent index of change is the slight, if statistically significant, fall in the mean number of activities per household, from 2.42 to 2.15. Part of this decrease may be a result of specialization within the household economy, of moving resources from less to more profitable endeavours. The data suggest, however, that the 'specialization' going on in many Milambo households is not strategic in the pro-active sense, but more a reflection of disappearing opportunities, and of declining capital resources. It should be kept in mind that these are very crude indices of social activity, which tell nothing about the volumes of commodities or cash involved. But even this indicator is enough to confirm the general impression of a contracting domestic economy. It is noteworthy that the number of households with no source of income, while still minute, has doubled.

Crop production

year	farmers	hectarage	production	sales
1979*	8	8.3	110	na
1980*	18	14.3	150	133
1981*	20	na	330	320
1982	51	84	2324	2188
1983	84	54.3**	2014	1729
1984	82	47.5**	3167	2873
1985	106	110.5	2490	2242
1986	85	102.5	2434	2311
1987	305	198.5	5423	4955
1988	429	268	7587	6722

*data missing for Mulumbi depot **data missing for Milambo depot

Figure 3. Maize production in Milambo Multipurpose Co-operative Society area, 1979-88. Sources: Milambo and Mulumbi Agricultural Camp notebooks; Milambo Multipurpose Co-operative Society records. All statistics for 1979–81 are underestimated due to missing data from the Mulumbi depot. Production and sales should be up to 500–600 bags higher. Production figures for 1983 and 1984 are estimated on the basis of sales volumes.

By the 1980s a commitment to cash-market oriented agriculture was gelling for the first time among Chief Milambo's subjects. In the Zambian context, cash-cropping was understood by all to mean hybrid maize production. This commitment was consummated in a remarkable maize 'boom' similar, if more modest, to that underway in neighbouring Northern Province from the mid-1970s (Sano 1989). In Milambo, over a ten-year period, maize production rose dramatically, from well under 1,000 ninety kilogram bags of white maize in 1979 to 7,587 bags in 1988.[6] Since more than 90% of the estimated production is officially marketed in this area, sales of maize through the co-operative or NAMBoard experienced a similar upward spurt. This trend is portrayed in the next figure.

Figure 3 reveals that the volume of maize sales increased by more than 300% between 1982 – the year Milambo Multipurpose Co-operative Society commenced marketing activities – and 1988. There are several reasons for this – most obviously, a sharp decline in cash income opportunities in the urban centres. In the eighties, migration to the mines ceased to be a reliable strategy for Milambo men to acquire cash. As the options narrowed maize farming began to look more attractive.

At the same time, the Department of Agriculture, with considerable

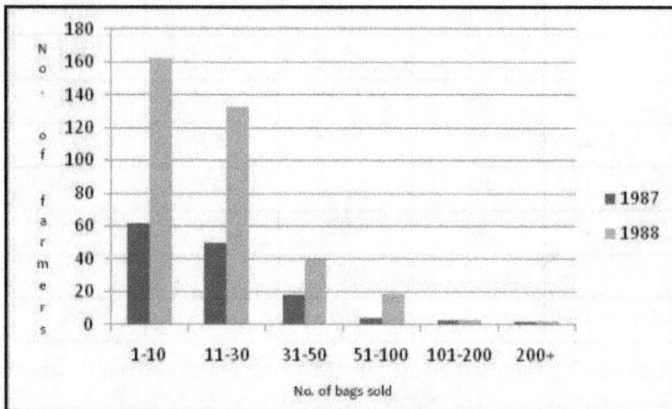

Figure 4. Numbers of farmers selling maize to MMCS in 1987 & 1988, by volume of sales.

donor support, was channelling its efforts into the 'Lima' campaign, targeted at semi-subsistence cultivators. The Lima Way involved the application of a modern package of hybrid seeds and chemical fertilizers to the production of any of five lima crops: maize, sunflower, rice, beans and groundnuts. From the beginning, maize has been the overwhelmingly predominant crop of choice. Recruitment for Finnida-supported lima training courses began in Milambo in 1980, with six farmers recruited for training annually from the Milambo Agricultural Camp area (Lwela and Mapula zones) for the next ten years.[7] By the time of the steepest upward swing in production after

---

[6] I have no data on production in Mulumbi in 1979. A realistic estimate might be 500 bags of maize bringing the society area total to roughly 600 bags.

[7] By the end of 1987, 40 (predominantly male) villagers in the Lwela-Mapula area had attended the lima training course. Of these 40 trainees, 21 farmers (all male) received credit through the Finnida CCS

1986, there was a lima farmer in virtually every village cluster in Milambo; the new technology was universally accessible.

As can be seen in Figure 4, growing numbers of semi-subsistence producers embarked on the path of lima farming at the height of the maize boom. Between 1987 and 1988, the number of farmers involved rose from 135 to 361.

The rising volumes of produce reaching the market were achieved through extending the total area of cultivation. Yields per hectare remain low, at both ends of the scale. The technology is simple to adopt, but difficult to master, especially under the prevailing conditions of erratic input supply.

	% of respondents planting crop	
Crop	1987	1995
groundnuts	41	75
maize	26	22
cassava	19	74
millet	16	50
beans	5	7

Figure 5. The proportion of farmers planting common crops in the 1986/87 and 1994/95 seasons.

In concert with lima training, the government launched an extensive Co-operative Credit Scheme (CCS) through the Zambia Co-operative Federation (ZCF). The base capital of the CCS was provided by donor organizations (Sida, Finnida). According to CCS principles, any reliable, healthy, creditworthy rural cultivator was entitled to a seasonal loan, in kind, of agricultural inputs for cultivating up to 2 hectares (8 limas) of land. In practice many factors (such as gender, locality, and education) combined to narrow the scope of access to CCS.

CCS loans also had several disadvantages for the small-scale producer. Success in submitting an application form demands literacy in English and the registration of guarantors in the case that the borrower defaults on the loan. Interest, at 18% in 1988, was far below market rates, but for the

(Co-operative Credit Scheme) in the 1987/88 season.

inexperienced farmer with low yields, even this amount of interest caused the producer's small profit margin to dwindle. The bureaucracy associated with the CCS scheme commonly resulted in borrowers receiving their inputs late in the planting season, thus reducing the probability of an optimum yield. Loan repayment processing delayed payment of the farmer's check. Finally, since eligibility for a CCS loan required co-operative membership, a CCS farmer was vulnerable to having society staff or leadership use the credit scheme to pry contributions to the capital fund or other forms of 'participation'.

The sudden upsurge in production after 1986 was increased further by a hefty price incentive to the producers. The 1987 price of K78 per bag of maize represented a 140% increase over the preceding year, and a 236% increase over that of the 1985 level. (This exponential rise levelled out in 1988 when the increment was only K2.)

Seen in terms of individual economic strategies, maize production was attractive to rural producers in two ways. First, it provided the most effective means of converting a local resource available in quantity – labour power – into a scarce commodity: cash.[8] Second, maize marketing could generate a single yearly check of relatively substantial size. With high inflation and little history with banking, there are many needs which can only be satisfied when a larger single sum of money is available (school fees, a bicycle, radio, blankets, shoes, large productive investments). Other forms of commerce – such as sales of fish, beer or other produce, road maintenance work, or even a monthly government salary – do not engender enough cash at a single time to allow for accumulated savings over and above daily consumption needs. The sporadic incomes from these various sources are too easily 'squandered' on daily requirements.

Although crucial to the dynamics of market development in Milambo, maize was not the crop most commonly taken to market – groundnuts was by far a more important commercial crop in terms of the number of households involved in its marketing, albeit not in terms of the volumes involved. Nor was maize the specialization of a distinct group of producers. Many if not most producers floated in and out of maize production, as the low repetition rate on CCS loans suggests.[9] When subsidies on inputs and marketing

---

[8] On the value of labour power and the 'cash squeeze' in a neighbouring Aushi community, see Allen (1998), as well as Gatter (1988).

[9] Only 40% of those CCS borrowers in 1987/88 who joined the co-operative society prior to 1986/87 were receiving credit for the second consecutive year. A proportion of the non-repeating farmers probably purchased inputs with cash, but I can't say precisely how many. Self-financing maize farmers consti-

were removed after 1992, the status of maize in the local production and marketing strategies continued to decline.

It must immediately be noted that there is a structural weakness in the data concerning groundnuts, cassava, millet and beans for 1987. Due to an oversight in the questionnaire, data concerning these crops was only recorded for families who reported having marketed a surplus of each respective crop. This omission does not affect the data on maize, since in 1987 virtually all maize production went to the market. Only negligible amounts of maize were retained for consumption since it made economic sense to sell one's maize to the co-operative and buy back milled maize at a subsidized price. The data collection problem was rectified in 1996. The latter data set is thus more comprehensive, and can be taken as providing a reasonable indication of the structure of production in the community.

Looking only at maize production, it is clear that the removal of subsidies undermined its attractiveness to Milambo producers considerably. No official data is complied concerning overall yields, but a comparison of the aggregate harvests reaped by survey respondents in 1987 and 1995 suggests that total output of maize is now about half what it was before liberalization. The number of maize farmers has fallen only slightly, but the volume of each farmer's production has dropped sharply. In 1987, fifty-eight households harvested 140 tons of maize (2.4 tons per farmer), while 8 years later forty-six maize-growing households brought in 63 tons (1.4 tons per farmer). Extensionists may be unhappy to learn that despite fifteen years of donor support to maize extension, average yields have declined from the already low level of 2,700 kg/ha in 1988 to 2,170 kg/ha in 1995.[10]

It is harder to get a handle on how groundnut production has changed over the period in question due to the recording error in the baseline data. Comparing those respondents who marketed groundnuts in 1988 and 1996, it appears that groundnuts have gained ground in Milambo farming systems. The proportion of the sample population now producing groundnuts for the market has risen from 41% in 1988 to 60% in 1996. The average amount of land such a farmer has under groundnut cultivation has increased slightly from 1.4 to 1.5 limas. Paradoxically, average output per producer has fallen over the same period, from 9.8 to 7.3 bags, while average yields per unit of land have also declined from 6.6 to 5.6 bags per lima.[11]

---

tute about a fifth of all maize producers.

[10] More on this later; improper seed may explain much of the decline in productivity.

[11] The unit of measurement we used for harvested groundnuts is a ordinary (so-called 90kg) grain bag filled with unshelled groundnuts. I take it that such a bag contains roughly 40 litres of shelled nuts.

This weakening of performance might reflect the large numbers of poorly capitalized, or inexperienced, producers entering the groundnut market for the first time. According to the 1996 data, market-oriented farmers produce, on the average, more than three times that of subsistence producers, and their yields per unit of land are more than twice as high.

These observations reinforce the qualitative impression one gets from talking with farmers. It is now 'common knowledge' that groundnuts have superseded maize as the cash crop of choice. The fact that donor interests and activities have now switched from maize to groundnuts is clearly a contributing factor. The Finnida-sponsored seed multiplication program (SMP) has been subsidizing groundnut production in Milambo for a couple of seasons, and just before the 1996 survey discussed here, SMP employees travelled through Milambo buying up groundnuts door-to-door at much inflated prices. To this extent, 'liberalization' has not yet achieved a perfectly free market situation. But there are other, less contingent, factors that contribute to the popularity of groundnuts in the present situation. First, groundnuts do not require chemical inputs. Seed can also be recycled from season to season. In addition, there is a functioning market for groundnuts, with the possibility of quite attractive margins, especially considering that the only cash outlays required are on the marketing side. I will return to these points below.

Groundnuts may have supplanted maize as the community's main cash crop, but maize still has its advocates. Thanks to the concerted efforts of the Ministry of Agriculture and Finnida throughout the 1980s, maize (mataba) has established itself as a core element in local farming systems. Maize is valued for a number of reasons, not least that people like the taste and it is vastly simpler than cassava to prepare in the rainy season. Therefore, there is a good local market for maize (sold by the tin), especially in the rainy season, and maize milling has also established itself as a form of local business in an economic terrain almost totally lacking in commercial nodes of activity. Having survived the vicissitudes of drought and structural adjustment, farmers are waking up to the idea that the inclusion of maize in their farming system lessens the risks of over-reliance on one commercial/food crop (that is, groundnuts). While there may be little economic justification for maize production on a more commercial scale in a marginal area like Milambo, small-scale (<1ha) stands are nonetheless beneficial to household food security, of the producer as well as to satisfy local demand. There remains however a residual nostalgia for the days of the maize boom, and some farmers continue to express a desire to resume semi-commercial (~10 ha) maize production.

Factors of Production

	Mean area under cultivation	
crop	1987 (n)	1995 (n)
millet	1.37 (35)	1.29 (57)
groundnuts	1.42(90)	1.51 (129)
cassava	1.71(42)	4.28 (47)
maize	3.84(58)	2.79 (35)

Figure 6. Mean areas (in limas) under common crops among surplus-marketing households (number of valid cases in parentheses).

Land Use
This section looks at changes in how farmers allocate their land to different crops, and in the overall amount of land taken under cultivation.[12] In terms of land under the cultivation of various crops, the economic reforms would seem to coincide with a number of changes. Comparing, as above, only those households in the two data sets marketing a crop surplus, the change in average area under the different crops appears in Figure 6.

Putting aside the inscrutable 1987 figure for cassava, the most dramatic changes in land use have affected maize.[13] Were one to extrapolate on the basis of these figures, it would appear that the aggregate area currently under maize in Milambo is only about 43% of what it was in 1987. By the same token, land under groundnuts would have increased by 65%. Such estimations are thumbnail sketches at best, but there is good reason to believe that Milambo farmers have reallocated former maize land to groundnuts and millet as well as, presumably, to cassava.

---

[12] Quite regrettably, it was not possible to gather quantitative information on the market in land. Qualitative data indicates that since land is generally abundant, the land market is very limited.
[13] I have no reasonable explanation for the low figures on cassava cultivation in 1987, but one must assume that cassava fields are grossly underestimated.

	1987	1995
Field type	Mean Area (n)	Mean area (n)
munda	1.5 (28)	2.6 (131)
chifuka	2.3 (42)	1.7 (99)
ibala	4.0 (19)	2.5 (143)
lima	3.5 (58)	2.5 (46)
ilungu	3.0 (1)	1.4 (3)

Figure 7. Mean areas under crops on different categories of field, 1987 and 1995.

	% farmers with distant fields	
Field by main crop	1987	1995
maize	2	nil
groundnuts	11	19
millet	17	25
cassava	12	26

Figure 8. Farmers whose most distant field is further than 90 minutes away

It is of interest to examine how land use is evolving in terms of the types of field farmers are utilizing. Farmers make sophisticated distinctions between different types of soil as well as between the types of plot they are using. We gathered information concerning five main types of field:

- *umunda* (a first year citemene plot, usually used for millet and groundnuts)
- *chifuka* (second and occasionally subsequent seasons after opening citemene plot)

- *ibala* (a 'permanent' field, as in one taken out of fallow)
- *lima* (a fertilized field, usually under maize)
- *ilungu* (a garden on the soggy rim of a dambo)

Respondent farmers had land under crops on these various types of fields as shown in Figure 7. The most striking information in this table concerns the run on umunda, or citemene land. Umunda land is land on which farmers make use of the nutrients embodied in the bio-mass of the trees and associated forest vegetation. The vegetation is cut and burned, and the resulting ash enriches the soil. Thus the increased amount of land appropriated as umunda can be connected to the decline in the availability of chemical fertilizers and the growth in groundnut production. In one way this development is counterintuitive, in as much as the death-knell has been sounded for citemene cultivation since early in this century (Moore & Vaughan 1994). From an environmental perspective, this trend may give grounds for concern. Trees are still abundant in the Milambo vicinity, but presumably similar trends are to be found in less remote areas of northern Zambia. Indeed, fuel wood collection and swiddening have already brought about deforestation at an alarming rate around Mansa and similar rural townships over the past decade. It is symptomatic that the number of respondents who report difficulties in obtaining land for citemene cultivation has more than doubled since 1988, increasing from 24% to more than half of the sample population.

This growing pressure on fresh, regenerated forest land for agricultural use is confirmed in Figure 8 by an examination of the distances that farmers must walk to their fields. Ibala or lima land suitable for fertilized maize production is in ready supply. But in order to find land conducive to the production of staple food cum new market crops, a growing number of farmers are forced to walk more than 90 minutes to their fields.

Credit

The availability of a subsidized credit/input package was a major factor contributing to the maize boom of the eighties. Correspondingly, the dismantling of this package goes a long way to explaining the decline of maize in local farming systems. The main problem confronting farmers who desire to produce and market maize is the difficulty is obtaining fertilizers (ammonium nitrate and so-called compound 'D') to counteract the innate nutrient-poverty of local soils. Fertilizer is expensive, and it is only available in Mansa, 80km away from Milambo. Without access to (seasonal) credit,

	1987 (n=41)	1995 (n=20)
% who applied for a seasonal loan	19	11
% granted a seasonal loan	17	7
Mean area for which credit was applied	4.6 limas	4.8 limas
Mean area for which credit was granted	4.4 limas	3.5 limas
Aggregate ground granted	182 limas	70 limas

Figure 9. Respondents' seasonal credit in 1986/87 and 1994/95.

few local producers can muster the necessary cash to purchase and trans-port fertilizers to their farms. As we have seen, the state-subsidized co-operative credit and marketing arrangements that the Zambian state pro-moted in the eighties responded – albeit imperfectly and inefficiently – to precisely this bottleneck. Inasmuch as the economic reforms terminated this arrangement, prospective maize farmers have been left in the lurch.

The policy of the MMD government has been to hand over the functions previously managed by the co-operatives – credit, input distribution and produce marketing – to private sector agents. In theory, enterprising pri-vate entrepreneurs will make the market function more efficiently than it did under the tutelage of clientelist and complacent parastatal agents. The drawback to the theory is that self-interested traders will only venture into a market where a reasonable margin is likely. In this respect – and especial-ly in the case of maize – Milambo does not appear particularly attractive.

In 1988, 40% of all respondents had applied for seasonal credit at least once in the past three years. In 1996, this frequency had fallen to 27%. The impact of the changes in the credit arrangements becomes very apparent when comparing the two seasons 1986/87 and 1994/95 in Figure 9.
Small-holders' access to seasonal credit has suffered a serious setback. This can be attributed quite directly to economic reforms. But how serious a problem is this and for whom? The groups most affected by the disintegra-tion of the state-sponsored agricultural credit system are those who would like to continue to pursue a career in maize production. Few, if any, ground-nut, cassava or millet farmers are interested in buying fertilizer on credit, and while there is no doubt a demand for capital to finance casual labour costs on a variety of crops, this has never been available through formal

channels. Persistent market-oriented maize farmers are a small group. But there is another, much larger group who would buy fertilizer for cash or on credit, if it were available for a reasonable price on the local market or in Mansa. These are the numerous farmers who would like to produce a small stand of maize for their personal consumption or for local sales, as well as those who have a line of vegetables or other crops with high soil nutrient demands.

Members of both of these groups have found themselves dealing with a new breed of 'credit coordinators.' These are private sector commercial operators who have been licensed by the state to work in the seasonal credit line. According to the conditions of seasonal credit provided during the 1995/96 planting season, farmers were to repay designated credit coordinators with 4 to 4.5 bags of maize per each lima of loan.[14] Given the average productivity of Milambo maize producers, the upshot of this is that the average producer in Milambo ends up with two to four bags of maize for each lima of loan, with the credit agency taking half to two-thirds of the harvest.[15] Under the prevailing system, which tends to bypass co-operative institutions, representatives of the credit agencies are obliged to deal with farmers on an individual basis. They thus justify the stiff conditions they offer on the basis of their high (transport) overheads.

One credit coordinator I interviewed cited 'typical' yields of 60 bags (5.4 tons) per hectare, or 15 bags per lima, as evidence of the capacity of local farmers to meet these conditions. Such management levels may prevail on more commercialized farms the likes of which have begun to emerge in close proximity to Mansa and other centres of demand. In a region like Milambo, reality is of another order entirely. Given the modest margins on maize production this distance from a major market, the private credit agencies are not making much more of a profit than the small-holders they are scalping. To some extent, these credit arrangements resemble the system of out-grower farming which is emerging in the more commercialized farming centres.[16] According to this system, the farmer becomes a de facto labourer of the credit agency who supplies the inputs and collects the crop, and who also determines the price at which the harvest will be valued.

---

[14] One lima = 0.25 ha, requiring 2 pockets of fertilizer (D-compound, ammonium nitrate) in conformance with established recommendations.

[15] The exploitative character of the seasonal credit facility is underscored by repeated complaints by Milambo farmers that no written agreements were drawn up at the time of the provision of inputs, while the repayment rates were only stipulated after the farmer had already harvested the crop.

[16] Outgrower schemes have been discussed by Chiwele et al (1997: 24).

At the pricing levels credit agencies have been employing in Milambo, it appears that all of the risk has been placed on the shoulders of the small-holder. Without a quantum leap in farm management levels, it seems unlikely that such a system can survive in this form.

Seed and Fertilizer

One of the aims of market liberalization is to enhance productivity by 'freeing up' factor markets. It is already clear that capital markets have not benefited from economic reforms – although the problem would seem to more one of Milambo's marginality to the market than of the market's failure to function properly. What is the situation with agricultural inputs? Under the co-operative marketing system, inputs were available in the locality at a reasonable cost – at least in principle. In practice, deliveries of seed and fertilizer often reached farmers weeks after optimal planting time. Does the new regime get the goods to the farmers any more effectively?

Maize. By the mid-1980s, the extension service of the Ministry of Agriculture had decided that pure hybrid maize varieties (SR52, SR752, R215) were too demanding for the climatic and management conditions prevailing in much of Luapula, and had begun to recommend the cultivation of the '600 series' of open pollinating varieties. These varieties did better under less disciplined management regimes. In addition, open-pollinating varieties can be replanted without the disastrous fall in yields typical of replanted hybrids. During the 1986/87 season, 90% of the Milambo maize farmers planted one of the 600 series of seed. In the vast majority of the cases (85%), the seed was obtained from a co-operative depot, in most cases the local MMCS offices. The bottleneck in the pre-reform market arrangements was the timing. Farmers are instructed to begin their planting with the onset of continuous rains. These typically begin in November. Yet of the maize farmers interviewed, only 15% had received their seed by the end of October.

In 1996, the situation was quite different. Two-thirds of the farmers continued to plant the open-pollinating seed, but now they were frequently (41%) drawing on their own reserves. Only one in ten obtained seed through a co-operative channel. At the same time, seed purchases from private sources had doubled (from 7% to 13%). Naturally, this self-help arrangement improved the timing of input supply. At least those two-fifths of the maize farmers who used their own seed planted on time. This wide-spread use of second (or higher) generation seed may be one cause of the fall in maize yields per unit of land noted above.[17]

Groundnuts. A number of agencies, including the Finnida-funded Luapula Livelihood Program, have been working for several years to improve the availability of high-quality groundnut seed through a subsidized seed multiplication project. Of late, the intention has been to focus on the development of local markets for improved seed. On the basis of the data available to me, this objective seems quite a way off. One of the few things that has not changed in the course of the market reforms is the propensity for groundnut farmers to rely on their own reserves. In fact, according to this survey data, the number of groundnut planters obtaining seed from a private source has fallen (from 21% to 17%) since 1988.

Labour

Human labour is the predominant, indeed almost the sole source of productive power in the Milambo domestic economy. Human muscle powering a hoe or an axe produces the bulk of the means of sustenance. Agricultural work is dictated by the seasons, and there are several peaks of demand for labour over the course of the annual cycle. The extent to which a production unit can meet the labour demands of these peak periods will determine the overall scale of household production. Thus an ambitious, aging or ailing household will seek to augment its own domestic labour resources with outside pairs of hands. The period of economic reforms has seen a considerable transformation of the local labour market. These changes can be characterized via a few key indicators.

Compared with 1987, twice as many households in 1995 employed casual labour in their farming activities. The number of labour-hiring households in our sample rose from 21% to 46%. Evidence suggests that this new demand for additional labour is emanating from the emerging groundnut sector. Surprisingly, perhaps, the absolute size of the labour force (approximated via the aggregate number of workers employed by respondent households) has hardly changed, nor has the number of workers hired by a household per unit of land under cultivation. The mean outlay per unit of labour increased 100 fold over the period 1988-96 (from 40 kwacha to K4, 000), whereas the value of a bag of maize is today 150 times greater than in 1988.[18] This could suggest that the supply of labour outstrips demand. On

---

[17] It would not seem to be a function of reduced fertilizer application. Compared with 1987, farmers in 1995 applied ten kilogram more fertilizer to a lima of land (averaging 84 kg/lima).

[18] A crude index estimated by dividing a household's total labour costs by the total number of labourers employed.

	% of farmers marketing crops by year	
crop	1987	1995
groundnuts	42	62
fish	22	25
maize	20	16
cassava	19	23
millet	15	27
fruit	13	6
vegetables	11	8

Figure 10. The percentage of all households marketing selected produce during the 1986/87 and 1994/95 agricultural seasons.

the other hand, respondents may be underestimating the value of labour costs paid in kind. During the maize boom, 89% of the transactions between labourer and employer involved cash. Today, payment is far less cash-based, with payment in kind becoming increasingly prevalent. Respondents may have calculated their labour costs on the basis of the value of the goods (salt, cooking oil, second-hand clothes) at point of purchase (Mansa, Copperbelt). In their transactions with labourers, however, the same goods will undoubtedly be imbued with a much-inflated rural value. In any event, liberalization has clearly extended the scope of the local labour market, and the evidence suggests that rising groundnut incomes are circulating within the local economy in the form of casual labour fees.

The Marketing System

Commodities

According to survey data, Figure 10 shows that Milambo households marketed surpluses of their primary produce before and after the economic reforms. This data reveals that in spite of heavy subsidies favouring maize marketing during the 1980s, only one-fifth of Milambo households sold surplus maize to the co-operative at the peak of the maize boom. While in bulk and value terms maize dominated the primary produce market, as a strategic income option, groundnuts were by far the most popular commodity.

This observation underscores the difficulty of speaking of 'the' market in rural Luapula. The rapid expansion of hybrid maize production and sales in the eighties did not occur in a commercial vacuum; nor did the establishment of a network of co-operative marketing depots for the extraction of maize to the urban sector constitute a transformation of agrarian trade as a whole. Brelsford's study of Copperbelt markets in the mid-1940s and Miracle's parallel analysis in 1959 both indicate the presence of Aushi traders on the line-of-rail, and while much of the Luapula trade revolved around fish, it is likely that the trade in agricultural produce between rural Mansa and the Copperbelt has a history as long as that of labour migration to the same destinations (Brelsford 1947, Miracle 1962).

Vigorous state intervention in the grain market from the seventies onward created a distinction between formal and informal trade. Formal

	groundnuts		millet		cassava		fish	
	1987	1995	1987	1995	1987	1995	1987	1995
Local Market	49	29	89	67	81	72	94	98
Traders	52	30	23	68	17	73	25	99
Outside Market	7	31	na	69	10	74	10	100

Figures 11. Percentage of households using various marketing channels for primary food crops in 1987 and 1995.

trade involved the moving of subsidized, license and price-controlled produce – above all maize. Parastatal and co-operative institutions had a monopoly on this. Informal trade covered everything else. The unlicensed trade in price-controlled produce also fell under the rubric of informal trade. One of the main 'informal' activities during this period was thus the border trade in maize meal from Zambia over the Luapula river into Shaba (Zaïre). According to the bureaucratic logic underlying this taxonomy, the small-scale Luapula-Copperbelt trade in primary produce such as cassava meal, fish and groundnuts also acquired the stigma of 'informality.'[19] The

---

[19] This attitude resonated in popular parlance as rural Aushi would refer to the small-scale private traders who came to buy or barter for groundnuts and cassava as 'banakungula' - the women who

UNIP government's agrarian policy thus imposed a nefarious normativity on the domestic economy. That was, only maize producers were 'real' farmers – everyone else was a 'semi-subsistence cultivator', while private traders (be these the producers themselves or outside middlemen) appeared as shady characters whose honesty and loyalty to the Party and its Government was clearly suspect.[20]

Spatial Configuration of the Market

If nothing else, trade liberalization has eliminated the normative distinction between formal and informal commerce. It has also led to a reconfiguration of how producers and buyers are linked in spatial and social terms. Figures 11 summarizes the shifts in the spatial relations of agricultural trade that have emerged in the course of Zambia's economic reforms.

A striking feature of this data is the increasing scope of private trader activity. The figures in the table tell nothing about the volumes of food moving through the respective channels. Nevertheless, it would appear that growing amounts of locally produced foodstuffs are sold to outside

Sites for sales of maize	% of farmers (n=37)	% of volume (n=344)
Local market	11	8
Traders	41	39
Outside market	16	20
Co-operative	24	33

Figure 12. Maize marketing in 1994/95 season. Proportion of farmers selling to various markets; and the proportion of volumes of maize (90 kg bags) marketed through same channels.

actors and being consumed elsewhere. The data also suggest that local producers are expanding their control on the marketing of fish and groundnuts – the most valuable commodities in both monetary and nutritional terms – by taking the produce themselves to an outside market (the District

---

clean out our groundnuts.

[20] This did not, however, dissuade leading politicians and civil servants from engaging in 'informal' trade - including the prescribed border trade in mealie-meal, cooking oil, salt, sugar and other basic commodities which were chronically in short supply.

bomas of Mansa and Samfya are typical destinations).

Together, these observations support the thesis of greater market integration. It should be kept in mind that the Luapula area has supplied the urban mining sector with food and labour since early in this century (Musambachime 1981). Without hard facts about the volumes of produce moving through the various channels, is it difficult to assess the extent to which the current situation represents a qualitative transformation of these age-old (informal) trade networks for the marketing of groundnuts, millet, fish and cassava to local towns and the Copperbelt. The perception of the farmers I spoke with was of a much intensified market in the 'informal' crops listed above following liberalization. Clearly, market reforms have opened up opportunities and stimulated the market for high value-for-bulk commodities – items for which the proportion of transportation and marketing costs per unit do not exhaust the profit margin. Figure 12 presents the structure of the current maize trade in Milambo.

Lacking data on 'informal' maize sales in 1987, it is not possible to present a similar comparison of the spatial shift of the maize market over time. In 1987, it was generally assumed that nearly all of the maize produced was marketed via the co-operative. The existence in those days of a 2-3 functioning hammer mills within the local community nevertheless indicates that a number of farmers were processing and consuming their own maize. Plausibly, some of this was also being traded on the local market. While I don't have quantitative data on the pre-liberalization local market, there is no doubt that the amounts of maize remaining within Milambo were minor, as is the case today.

On the basis of this survey, talk of the total demise of the co-operative institution seems premature. In 1994/95, the local co-operative society shared the costs and benefits of marketing maize on nearly equal terms with the traders. What is most important to note in this connection is that economic reforms have not really altered the spatial configuration of the maize market. Maize is still predominantly an 'export' crop. What has changed is the gallery of actors involved in the maize trade. Co-operative members are now a small minority vying with unorganized individual farmers for trade outlets; the one or two marketing employees of the primary co-operative society compete with a fluid throng of private traders and 'credit coordinators' for local produce. Another point is that while the volume of maize remaining in Milambo is minor, farmers themselves see local maize sales as a growth industry. Inasmuch as the groundnuts continue to be in high demand as a cash crop, it seems reasonable to assume that local consumers will have a growing capacity to buy maize. This trend can also affect the emerging social relations of the market in agricultural produce if and when clear specializations emerge within the domestic economy.

Social Configuration of the Market

Actors and Transactions

In terms of their social content, economic transactions can be personal or impersonal. Trading within the community has a strong personal element;

	% Responding with answer		
crop	1–2 buyers	Many buyers	n
maize	26	57	35
groundnuts	45	46	132
millet	11	72	55
cassava	26	58	47

Figure 13. Number of marketing transactions for various crops in 1995. Respondents were asked to recall the number of customers for each crop. They either provided a discrete number, or answered 'many.' Table 12 reports the number of respondents giving discrete numbers of 1 or 2, or responding many. The residual respondents gave answers fluctuating between 3 and 9. Rows do not total 100% due to the incidence of 'don't know' responses.

the relationship between the buyer and the seller will generally have dimensions beyond the immediate transaction at hand (involving clan/lineage membership, or residential locality). The terms of transactions are often shaped by these social factors. One may offer millet to a neighbour who must brew katubi in order to attract labour for thatching her home on terms not dictated directly by the prevailing market situation. Selling to a trader from outside the community, or taking produce to an external market generally involves more impersonal relationships. Personal histories are not as likely to impinge on the terms of the transaction. From the above discussion of the spatial relations of the market, one can induce that impersonal transactions are common in agricultural trade. Irrespective of the crop involved, most of the dealings people have are with outside traders or at a market outside the community. Does the social context of a transaction affect returns to labour and the economic security of the small-holder producers? How do the farmers price their commodities? Do producers establish stable marketing arrangements with particular clients? Does the impersonality of a deal affect the producers' ability to negotiate a reasonable price for buyers?

From the Figure 13, it is evident that there is no 'typical' type of transaction in the food crop market. Looking at the number of customers to which

	Mean wholesale price	n	Mean retail price	n
Local Market	1690	7	1570	22
Traders	1620	21	1580	47
Outside Market	1510	6	2000	24

Figure 14. Average price obtained for groundnuts (5 litre tin) through different market outlets (1995).

producers sold their produce in 1995, two polar groups emerge: those that sell their surplus to one or two customers and those who market to a large number of buyers.

This evidence indicates that small-scale, sporadic transactions dominate the market. The impression is that producers sell small quantities of produce as the need for cash arises; or as something attractive is offered by a travelling trader for barter. This may corresponds to the whims or resources of the customers, but a more plausible reading of the data is that self-marketing producers prefer to sell their produce over a longer period of time. The main reason for this is the wildly fluctuating price differentials for most produce over the course of a marketing season. Immediate post-harvest prices in May and June can easily be a tenth of what they will be during the 'hungry' season of February-March when the previous crop has been sold or consumed and the new harvest is not yet ripe. Producers are under heavy pressure to sell off some of their crop as soon as it is harvested in order to meet short term loans and seasonal expenses such as school fees.

After the most pressing obligations are met, farmers with the means to do so will hold back as much of their crop as possible in anticipation of peak prices. It is a risky business however; one never knows which passing trader will be the last, or whether an early rain will cause unprotected grain to rot.

How does the marketing outlet and type of transaction affect prices? Figure 14 summarizes 1995 price data for groundnuts when sold via different market channels. A distinction is made between wholesaling (to 1 or 2 customers) versus retail sales (to many customers).

Predictably, producers fetch the best mean price for their groundnuts when retailing at an urban market. Transaction costs are also highest under such circumstances, since the retail sale of any larger quantity of groundnuts in 5 litre tins can require sitting in the market for days on end, in addition to transportation costs for both produce and marketeer. Bulk sales on the local market entail the lowest transaction costs; this form of marketing also brought in the second best average price. The opportunity is probably rare, or involves hidden costs, since very few of our respondents sold their produce in this way.

Do 'primordial,' personalistic social relations constrain market development? There is no direct evidence to support such a claim. The obstacles to 'modernization' that Hyden and others have seen emanating from a primordial 'economy of affection' would seem to be tied more to the 'modern' parastatal sector, than to the 'traditional' realm of consanguinity and locality. The view one gets of Milambo in the Third Republic is of an economy in transition and under immense pressure to adapt to uncertain and fluid conditions. Farmers are seizing new opportunities within the limitations of their resource base and of the information available to them. We have had a glimpse of how social relations are undergoing a transformation on the level of the regional market. This transition has yet another dimension, however, within the domestic economy and its basic economic unit of the household. These issues are the focus of the next section.

The Impact of Economic Reforms

Changing Intra-Household Relations

Our evidence suggests that the economic reforms have had a tangible impact on rural households, not only on how individuals relate to the market, but on how spouses within households relate to one another. What

	1987 (n=189)	1995 (n=192)
Men control marketing	60	40
Women control marketing	40	60

Figure 15. Control of crop marketing by gender.

Who controls maize marketing?		
	1987 % (n=55)	1995 % (n=32)
Male head	89	66
Female head	7	6
Female spouse	1	19
Spouses jointly	nil	9

Figure 16a. Maize.

Who controls groundnut marketing?		
	1987 % (n=85)	1995 % (n=122)
Male head	42	36
Female head	12	7
Female spouse	31	26
Spouses jointly	15	31

Figure 16b. Groundnuts.

Who controls millet marketing?		
	1987 % (n=33)	1995 % (n=52)
Male head	33	14
Female head	3	19
Female spouse	52	67
Spouses jointly	12	17

Figure 16c. Millet.

Who controls cassava marketing?		
	1987 % (n=41)	1995 % (n=43)
Male head	44	16
Female head	7	7
Female spouse	29	40
Spouses jointly	20	37

Figure 16d. Cassava.

we are seeing, I propose, are new, even intensified forms of intra-household competition, but also a nascent trend toward shared economic strategies among family members. This can be seen rather clearly via an inspection of the shifting division of labour between spouses in the marketing of surpluses of primary produce. These trends are naturally emerging against the backdrop of the major shifts in production and agrarian commerce discussed above: the relative decline of maize in favour of groundnuts as an export cash crop, reflecting an overall broadening of the opportunities for marketing traditional food crops like millet and cassava. This shift from the market domination of maize, a 'male' crop, to groundnuts, associated with the female realm, is reflected in a comparable shift in the role of men and women in crop marketing.

As revealed in Figure 15, which aggregates gender specific data on the control of maize, groundnut, finger millet and cassava marketing, male and female positions have flip-flopped during the course of market liberalization. Not reflected in this table is a second very tangible trend – the rise of collaboration among spouses in the marketing of crop surpluses. This will be taken up in more detail further on.

Taken crop by crop, one can see how the gendered shift in marketing responsibilities (or resource control) has occurred. For maize, the conditions reflected in Figure 16a signal a vigorous rise in the role of women in crop marketing. Eight years ago it was unheard of for somebody's wife to market the family's maize. Today, this occurs in one household in five, and in another ten percent of the households, men and women market their maize jointly.

Several factors are at work here: a decline in the economics of maize production has perhaps led men to 'abandon' the crop to their wives, while in some cases women's attraction to maize reflects its rising status as a wet-season staple. The contrast with groundnut marketing is once again revealing.

For one thing, this table brings out the substantial role that men played in groundnut marketing during the maize boom, prior to the economic reforms of the nineties. In popular parlance, as well as in the regional literature, groundnuts have been considered as predominantly an informal and hence a 'female' crop. Yet in two-fifths of the households, men controlled the proceeds from groundnut marketing in the pre-reform period, putting male and female control at roughly equal par. This is an indication of how forcefully the androcentric emphases of the maize promotion policy (which favoured men in lima training, co-operative membership and credit access) thrust men into the forefront of household marketing functions. Since the reforms, the role of individual spouses in the groundnut trade – both men and women – has declined, to the benefit of joint household control of groundnuts. The popularity of this marketing arrangement has doubled since 1987. As we have seen, the groundnut trade is currently where the commercial action is. Groundnuts offer an attractive source of direct income with a low demand for cash inputs. Groundnuts also represent the latest possibility for benefiting from a donor supported program, i.e., via contracting out as a seed producer. This may explain why exclusively female control of groundnuts has been weaker than with other crops. Yet compared with the total sidelining that women experienced during the maize boom, they have held their own rather remarkably in the bullish groundnut market. With respect to the millet trade, women have fared even better.

A major difference between groundnuts and millet as an item of commerce is that Milambo farmers very seldom take their millet to an outside market. As we have seen (Figure 11), visiting traders are taking a larger share of the millet harvest. Yet for the most part, millet remains within the community where it is the most important raw material for traditional opaque beers. For many single women (widows and divorcees) who lack the labour resources to produce a marketable crop surplus, beer brewing is their main source of income. This interest in the crop is reflected in the

Number of crops marketed jointly	1987 % (n=221)	1995 % (n=214)
0	91	75
1	6	18
2	2	5
3	0.5	1.5

Figure 17. Proportion of households in which spouses market a crop together.

strong growth in female control of the trade in millet, and lends credibility to the hypothesis considered above that the reforms have catalyzed a degree of specialization within rural localities.

Cassava, once exported in quantities to Elizabethville (Lubumbashi) and the Zambian Copperbelt, is again the object of growing commercial interest. Cassava meal can also be used for beer, but its main use is for preparing ubwali, the staple food. Urban demand for a cheaper alternative to maize mealie-meal has no doubt been the motor force behind trader activity on the cassava market. Looking at the final table in this series, it is evident that men have relinquished (or surrendered) their former domination of cassava marketing to their spouses. In a significant number of cases this has led, again, to spouses sharing of the responsibilities and plausibly the proceeds of marketing the crop. Such collaboration was more prevalent in cassava marketing than with other crops in the pre-reform period, and this continues to be the case. The parameters of this unmistakable trend toward intra-household co-operation have received scant attention in the literature and deserve a moment's consideration.

Corporate Households?

The notion of the household has been the target of intense debate within agricultural economics for more than a decade. Feminist scholarship especially has staunchly criticized the 'household model' of the rural economy, arguing that rural African households do not conform to the corporate decision-making ideal of neo-classical economics. Indeed, many writers reject the household concept altogether in favour of a strictly gendered perspective on economic behaviour. According to this view, one should conceive of rural farmers as individuals and study their actions in terms of a struggle for the control of resources along gender lines.[21]

There is much to be said for the feminist critique of neo-classical economics. Yet a position which rejects the corporate notion of household altogether misses two inter-related points: one, there are a number of different kinds of 'households.' In Milambo, for example, one indeed finds those in which men and women wage a gender war over the allocation of land and labour, as well as over control of the proceeds. But there are also households in which husband and wife work hard at collaborating for the common good of themselves and their offspring. This highlights the second point, that from a 'developmental' perspective, it is often those families attempting to function as a corporate entity that manage to avoid many risks and capitalize on opportunities arising in the economic environment.

These aggregate data are a coarse indicator. The unit of analysis is the household and not volumes of surplus, so it is not possible to estimate the amounts of economic value represented by the three categories of male, female and joint control of resources. Nevertheless, these tentative observations point toward an important area for further research. It is also important to identify the factors which can hinder or facilitate the reconfiguration of intra-household arrangements. An anecdote from my

---

[21] Moore (1994) and Moore & Vaughan in Geisler, (1992:113-39).

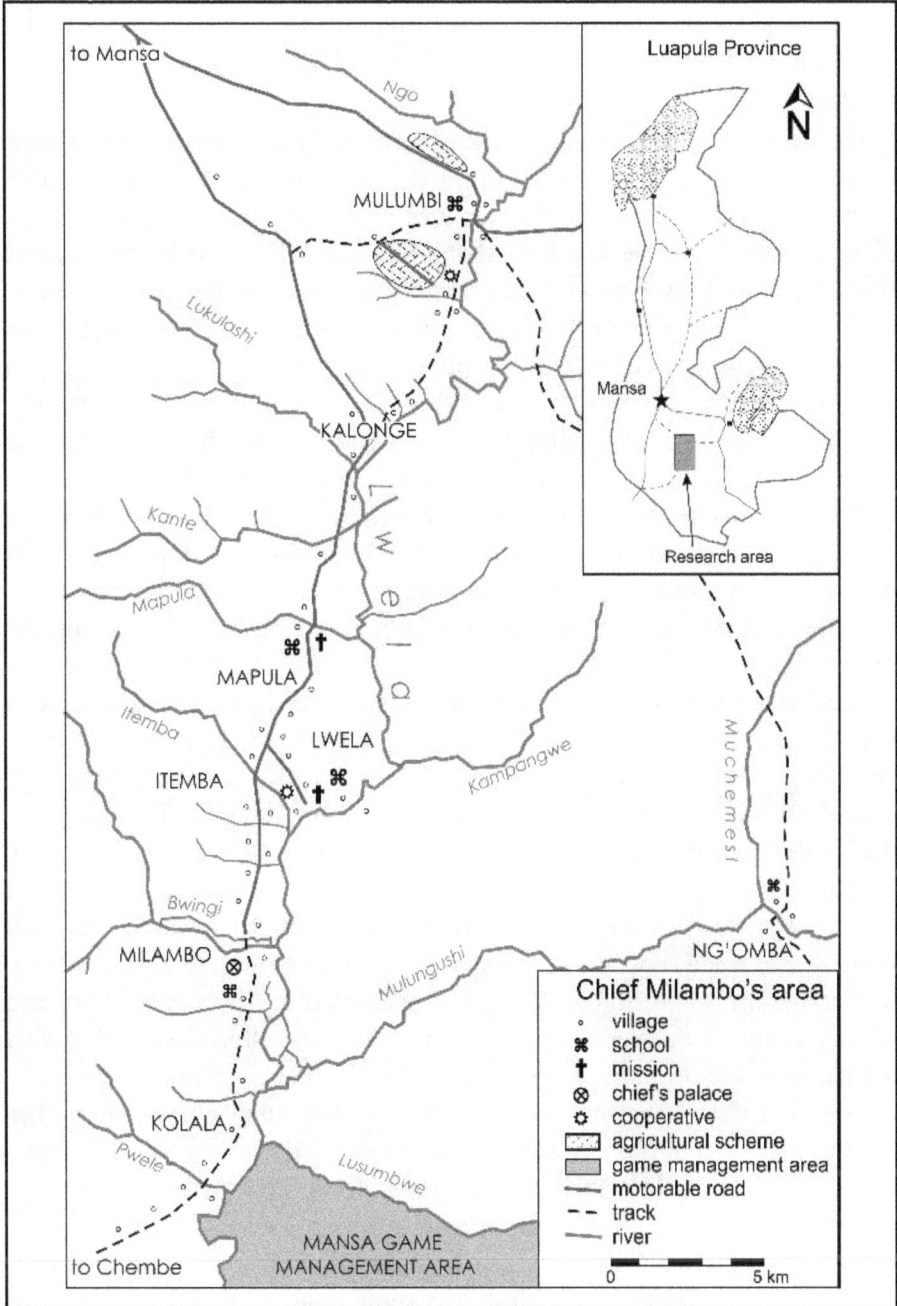

Figure 18: Chief Milambo's area.

fieldwork illustrates of how a seemingly spurious factor like the state of the roads can be decisive in shaping economic relations within the household.

Roads and Equity

As we have seen, the groundnut trade has traditionally brought significant flows of cash (and barter goods) into the local economy. In the past, much of this flow had been managed by female household members. In the course of my fieldwork in 1996, I undertook to investigate the paleo-feminist thesis that men would respond to the expanding cash market for groundnuts by attempting to wrest control of groundnut marketing away from their wives and for themselves. To test this hypothesis, I organized four meetings with PEAR (Participatory Extension and Research) and comparable community groups along the Lwela valley to discuss the changes in the marketing system and the current situation as regards the groundnut trade.[22]

The first meeting was held in Kanyensha about 6km south of the Chief's palace in the area known as Kolala. Three PEAR groups (including a seed multiplication group) have been operating in Kolala since before the previous (1995) season, and their chairperson (a man I got to know during my previous visit as the sub-depot chairman for the Milambo co-operative society) is an exemplary leader who has succeeded in giving the groups both form and direction. Our meeting was well attended by about 30 farmers. There was a lively discussion in which both men and women took part actively. I was not surprised to find the women complaining that their husbands had taken over the marketing of the household's groundnuts. 'My husband,' said one young woman, 'takes our produce to Mansa for sale, and only brings me back one citenge!' When asked why this was so, the men replied that since they were stronger, it was naturally their responsibility to manage the marketing exercise. This seemed to be such an obvious case of male chauvinist domination, that I found myself ending the meeting with a discussion about the need for co-operation in the household.

The next three meetings took us progressively further north along the road to Mansa: to Milambo, Meleki and Mapula. To my surprise and confusion, the situation described by the women (and their husbands) in these communities was exactly the opposite of what we were told in Kolala. In all three of these latter meetings, farmers assured us that most often women continue to market the groundnuts. In some cases men and women both

---

[22] The formation of community-based PEAR groups has been vigorously encouraged under the auspices of the Luapula Livelihood Programme sponsored by Finnida in a number of communities including Milambo.

cultivate and market their crops individually. In many instances spouses undertake the production and marketing activities in joint collaboration with one another. Such was not the case in 1988. At that time it was a rare household in which the male 'farmer' would sit to discuss with his 'cassava cultivating' wife, how the family should use the proceeds from the maize harvest. It was assumed that this was the domain of the man.

On the basis of three of our four meetings, it seemed plausible to claim that market liberalization has had an empowering effect on Milambo women. By virtue of the groundnut boom, women have greater access to cash and other resources, and greater control over how household resources are utilized. The Kolala story remained an enigma, however. Why was it that in this, perhaps the most active and organized of the Milambo communities, women have failed to benefit as much as their sisters in neighbouring villages? Contemplating the map (see Figure 18), an answer suggests itself. The main factor that distinguishes Kanyensha from the other communities in question is the fact that 'Milambo transport' never reaches Kolala. The terminus is at Milambo, and the 'lifts' often stop north of Bwingi stream. The main reason for this is the condition of the road. After Milambo's palace, the road to Kasomalwela rapidly degenerates into a winding and dipping bush track, a definite risk for most of the ageing vanettes that make up the Milambo transport fleet. For a Kolala farmer to market her produce requires dragging one or more 80 kg bags of groundnuts 6-8 kilometres up the track to Milambo. 'Strength' is definitely an asset in this exercise, and it not surprising that men find themselves taking control of the task.

A more general point to which this story alludes concerns the definition of market liberalization itself. Contrary to the claims of Geisler and others, the suffering of rural women has not necessarily deepened due to market liberalization. Rather the opposite seems to be true – at least in this specific case. In the Kolala anecdote, the failure of women to benefit more fully from the groundnut trade – and of households to react to the new market opportunities as a corporate entity – derives from the fact that, for the women, the market has not really begun to function. Kolala women are not free to market their crops to their best advantage because the poor condition of the road obstructs free market access. The most immediate lesson from this incident is simple. If one wants to enhance the capacity of the rural poor, both men and women, to benefit from new economic opportunities, make sure that the basic infrastructure is in place. The shortest path to greater gender equality would seem to travel along a well-graded road.

It appears that families in neighbourhoods served by a functioning transport network were able to take advantage of the benefits of collaborative marketing. The extent of these shifts in intra-household relations should be quantified; but even more important would be to investigate the practical implications of collaborative marketing arrangements for the ideology (or normative basis) of gender roles in rural Zambian society. This, one assumes, will have important consequences for household economic strategies, capital accumulation, and social issues like child nutrition and welfare.

Social Differentiation

It is obvious to the naked eye that there are considerable discrepancies in the resources embodied in different Milambo households. Houses with corrugated iron roofs, hinged doors, pane glass windows and cement verandahs coexist with thatched huts of poles and mud. A few compounds sport evidence of major capital investments – a (dilapidated) Land Rover or a hammer-mill – and a number of prominent sitting rooms display factory-

Quintile	Representing total assets less than (ZMK)	Mean assets within quintile (ZMK)	% of community assets	n
(poorest) QI	220	120	0.9	44
QII	752	480	3.6	44
QIII	1,710	1,780	8.9	44
QIV	3,044	2,310	17.5	44
(richest) QV	108,318	8,910	69.0	45

Figure 19. Stratification in Milambo, 1987. Cut-off points, mean assets and shares in overall community assets controlled by each quintile (20%) of the population. Source: 1988 survey.

quality sets of furniture. But the surface appearance of things is not the whole truth. Fear of their neighbours' jealousy and of the inevitable accusations of sorcery constrains people's demonstrativeness. By the same token, an impressive front can prove to be the mere shell of past wealth long since dissipated.

One of the objectives of the socio-economic surveys carried out in 1988 and 1996 was to gather data on household resources in order to quantify the nature and scope of the differences between social strata. Needless to say, the reliability of confidential information revealed to an anonymous passer-by has its limitations. Even when willing to confide in an enumerator concerning delicate matters like income, credit relations, production levels and financial affairs in general, people do not always remember details accurately. Aware of these methodological shortcomings, we tried to keep the interview design focused on practical matters of recent occurrence. The data constitute a critical and reflexive attempt at surveying the household economies of a relatively good sample of the Milambo population. Still, conclusions and extrapolations on the basis of the ensuing analysis must be approached with utmost care. Still, within the context of these limitations these data provide the only quantitative overview available of recent stratification trends in rural Luapula.

Pre-Reform Social Structure

To measure the resource endowments of each household, information was gathered concerning both fixed and liquid assets.[23] Households were then ranked into quintiles of the population according to their estimated worth in kwacha terms.[24] It will be helpful to begin the analysis with a summary of social differentiation trends during the maize boom years. The overall distribution of wealth in Milambo in 1988 is presented in Figure 19, which gives the cut-off levels and basic characteristics for each quintile, or fifth of the population.

The table speaks graphically about economic discrepancies within the community at the crest of the maize boom. The richest fifth of the population controlled more than two-thirds of the productive wealth, while the poorest twenty percent represented less than one hundredth of the sum of all household assets in the area. The mean worth of a household in the uppermost quintile was nearly 75 times that of a family on the lowest rung of the ladder.

It would be interesting to know more about the historical trends behind

---

[23] The variables included in the measure of wealth include the value of capital goods (agricultural and other implements, transport, livestock, fishing gear) as well as income flows from wages, the sales of produce and credit. It should be stressed that the final sum arrived at is at best an indicator, and not a comprehensive inventory of wealth. Not included in the measure are the value of household structures, furniture, kitchenware and other household appliances, clothing and holdings outside the village milieu.

[24] Each quintile represents 20% of the population.

these discrepancies. Was this a new phenomenon, or was this degree of stratification a feature of Milambo social structure even before the maize boom? With no comparative data from an earlier period it is difficult to make any conclusive interpretations. Marked stratification has certainly existed in rural Luapula at least since circular labour migration in the urban sphere introduced differential access to cash resources (Cliffe 1978: 327-8; Bates 1976; Pottier 1988). What distinguishes the 1980s from earlier processes of stratification in a qualitative sense is the fact that the influx of maize incomes created a structure of endowments grounded firmly in locally generated assets, that is, on a spatially unified process of agricultural production combining local land and labour.

One can note another factor which suggests that the eighties marked a radical break with the past. If we examine the relation of a household's fixed assets (the wealth invested in capital goods, like tools) to its total wealth (including liquid assets, the bulk of which accrued from crop marketing), it becomes clear that the richer the household, the greater proportion of its assets are liquid. Thus 90% of the families in the poorest quintile, QI, had virtually all their wealth tied up in fixed assets, while up in the richest QV, two-thirds of the households have more liquid assets than fixed. This is because the new opportunities for income from crop sales explain the greater part of the stratification observable through the survey data.

In an agrarian economy like Milambo's, the accumulation of liquid assets shows a great deal of fluctuation from year to year. This implies a high level of mobility between asset strata (i.e., quintiles) on a seasonal basis as people redefine their strategic priorities on the basis of changes in their resource situation, social obligations and perceived opportunities. In this connection it is pertinent to recall the earlier observation that there was roughly a 40% repetition rate for seasonal agricultural loans, indicating that liquid assets in the form of maize income were likely to vary greatly from year to year. This being so, any given household can bounce among the asset strata rather dramatically. One should therefore be cautious not to imbue the concept of 'stratification' with connotations of structural continuity and persistence. It is reasonable to assume that a certain core group of affluent households will manage to maintain their position in the upper economic strata of the community from year to year. It even more probable that resource impoverishment is a relatively persistent condition. However, evidence points to the fluidity and porousness of the social structure in

Milambo. Or as Sally Falk Moore has remarked, in rural Africa 'no condition is permanent.'

Post-Reform Social Structure

From the above it is evident that market reforms have had a marked impact on the domestic economy in Milambo. An attempt to quantify the effects of these reforms on household endowments on the basis of the available data collides with methodological difficulties. The main source of problems in comparing pre- and post-reform asset endowments derives from the relative growth in the 'informality' of the liberalized rural economy. Commerce in the era of centralized credit, input supply and marketing institutions occurred between farmers and formal institutions. Trade in the liberalized economy, in contrast, occurs predominantly between individual actors, with a multitude of small transactions dominating the commercial realm.

Few if any records are kept of these transactions. In addition, a growing proportion of exchange in the domestic economy involves barter and reciprocal trade, in which cash currency is only marginally involved. Even were non-monetized transactions on the decline in absolute terms, their weight in the economy as a whole is probably greater today than in 1987, when formal cash transactions in maize accounted for the bulk of the cash in circulation in Milambo. At that time it was possible to cross-check the aggregate volumes of produce and sales as reported via survey data against the official records of the extension and marketing institutions. Now these

Quintile	Representing total assets less than (ZMK)	Mean assets within quintile (ZMK)	% of community assets	n
(poorest) QI	61000	32210	2.5	43
QII	142500	99070	8	44
QIII	239700	193130	15	42
QIV	360500	294920	22	42
(richest) QV	1774000	691960	53	43

Figure 20. Stratification in Milambo 1995. Cut-off points, mean assets and shares in overall community assets controlled by each quintile (20%) of the population. Source: 1996 survey.

official registers are of little use, and the researcher is completely dependent on what individual respondents can recall and are willing to report about their incomes.

Another problem hampering the comparison of assets is inflation. Between 1987 and 1995, the Zambian kwacha has deteriorated at a devastating rate, from about 9.5 kwacha to the US$ to roughly 860:1, a 90-fold devaluation (EIU 1996). The extreme rate of inflation can clearly be inferred from

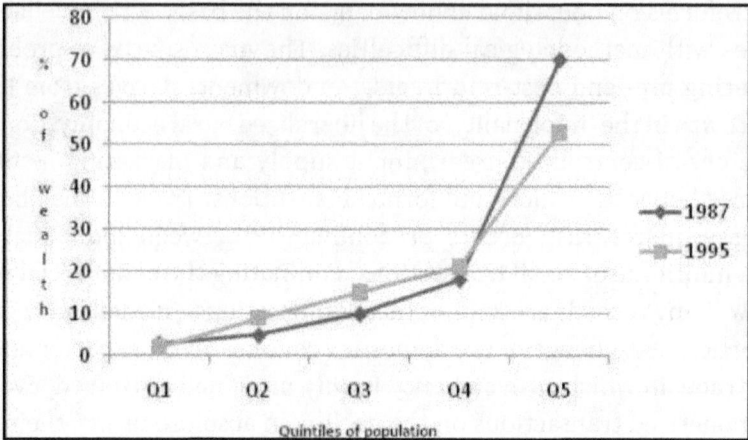

Figure 21. Control of aggregate wealth by quintiles of the population in 1987 and 1995.

the data in Figure 20, which presents endowments by quintiles of the population in 1995 calculated on the basis of the same parameters used for Figure 20 above. Yet in rural Milambo, the exchange rate of the kwacha to the dollar does not convey the whole truth about the value of money. Much of the economy is only partially monetized, and an even greater part involves commodities (labour, locally manufactured tools, fish, beer, and so on), the value of which is scarcely affected by foreign exchange rate fluctuations. Indeed, the 'liberalization' of trade has also meant greater leeway to negotiate the value of money. With fewer households benefiting from public sector incomes or state-guaranteed floor prices for crops, inflation has affected different classes of goods very differently. Instead, then, of using an 'official' rate of inflation, we have chosen to inflate unknown values in relation to the price of maize on the open market.[25] The reasons for this are that maize is a commodity most commonly bought and sold for cash, and

---

[25] An 'official' rate of inflation might be put in terms of the foreign exchange rate or the Central Statistical Office's 'basket of commodities' estimates. By 'unknown values' we mean the costs of certain assets (fish nets are one example) for which the precise 1996 value is not known, but which are included in the overall 'wealth' index in order to allow for comparability with the 1988 data.

because market price data was readily available from a variety of sources. Using this index, it appears that by 1995, the value of the kwacha had deteriorated to about 150th of its worth in 1987.

Income Distribution

It is reasonable to assume that the calculations presented here underestimate household incomes accruing from local petty commerce in items like beer, fish, and crafts. At best, they give a heuristic overview of changes in the distribution of agricultural incomes. This is certainly not the full picture, but it should provide some insights into the impact of liberalization on an agrarian community like Milambo. Constructing an index of household wealth similar to that presented above for the 1987 data, income distribution in post-reform Milambo appears in Figure 20.

Two features of this distribution immediately catch the eye. One is the overall flattening of the distribution curve, indicating that income differentials between the richer and poorer citizens have become less steep. Figure 21 brings this out clearly. Closely related to the reduction of income discrepancies is the marked growth of the relative share of aggregate assets controlled by the 'middle' strata (quintiles II through IV). According to these data, the 'middle classes' (constituting 60% of the population) increased its share of aggregate assets from 30% in 1987 to 45% in 1995.

What one makes of these figures will depend on the frame of reference. It is possible to argue that market reforms have reduced social differences and made of Milambo a more egalitarian society. It is equally possible to claim that the retraction of maize money has simply led to a more even distribution of endemic poverty. From the perspective of classical modernization theory, liberalization would appear to have disrupted the healthy process of accumulation (which may have promoted investments in land and labour productivity), which had been launched (albeit structurally constrained) by the UNIP regime.

The figures themselves are not conclusive, and one must allude to a number of qualitative factors when judging these trends. One point to consider is the apparent expansion of the labour market. Our data indicate a greater use of hired labour. This suggests the improved redistribution and/or local circulation of incomes within the domestic economy. This can also point to emerging specializations within the domestic economy, which may be seen, depending on one's perspective, as the first step towards improving productivity, or as the emergence of antagonistic class relations.

We have also noted the relative empowerment of women in the commercial sphere. In feminist terms, this would be seen as a positive end in its own right. Seen from a developmental perspective, greater gender equality may also lay the foundation for the strengthening of the conjugal household unit as an economic enterprise – a trend which can also promote long-term accumulation and improved productivity. How this eventuality relates to the future of matrilineal descent and/or inheritance (which may or may not be a precondition for female control of assets) is beyond the scope of the present data. In any event, all of these elements are 'emerging' trends that deserve more systematic attention than I have been able to render here.

Concluding remarks

1. These results have identified a number of outcomes for the marginally situated small-holders in one corner of rural Luapula connected to the economic reforms of Zambia's Third Republic. According to our data, the reforms have produced both winners and losers, and their developmental effect can only be assessed against the long term trends in the domestic economy as a whole. The main issue in the long-term frame is whether market liberalization will lead to rising incomes, accumulation and investments in productivity. There is no doubt that to some extent liberalization has freed up some developmental momentum, as its advocates submit. It is not clear, however, to what extent this momentum is leading to sustainable outcomes. The problem is of the lack of an overall policy framework which would harmonize macro and micro trends. An important micro-trend is of growing agro-entrepreneurialism and the more equal participation by women in the market as well as in conjugal household decision making. The larger picture is of local producers having an increasingly difficult time securing adequate economic services (inputs, credit, and marketing). Closely related to this is the deteriorating quality and availability of the basic public goods necessary to support a decent life in a rural community like Milambo.

In marginal areas like Milambo, 'liberalization' has meant that responsibility for the establishment and maintenance of health, education and economic infrastructure is increasingly being relegated to donors and the 'voluntary' sector. What is worrisome here is that the smaller bilateral donors and third sector agencies represent much splintered interests, and tend to pursue piecemeal, 'social engineering' solutions to what are fundamentally basic economic problems. The currently fashionable emphasis on forming and supporting informal 'groups' of women, youth and other 'vulnerable' sections of the population is guided by the premise that (women) farmers need to be socially empowered in order to take advantage of the emerging market opportunities. At the same time, larger donors (like the World Bank, the EU and USAID) are concentrating their efforts on 'governance' issues. In the agrarian sector this has meant a focus on reforming agricultural administration (under the auspices of the Agricultural Sector Support Program) in the hope that this will improve the financial management of donor investments.

Neither of these approaches gives anywhere near adequate attention to the fundamental issues of labour and soil productivity. Indeed, looking back over the past 20 years of intervention in Luapula agriculture, the only sustained endeavour which has made lasting gains in improving the lot of small-holders has been that of the Adaptive Research Planning Team. The achievements of the ARPT have been the result of painstaking work on identifying and improving adaptive varieties of crops with a proven track record in Luapula conditions. After two decades of systematic work – based increasingly on research collaboration with local farmers –ARPT has been able to release much improved varieties of many crops (e.g., cassava, groundnuts, millet, sorghum and rice) to local farmers. Regrettably, an effective means of disseminating this improved genetic material to the farming population has, to my knowledge, yet to be established.

At the same time, the long-term deterioration of the public economy (more a cause than an outcome of the economic reforms) has undermined the very basis of modern welfare in much of rural Luapula. While the local clinic and Basic School at Lwela, for example, continue to function, there is no doubt that their financial basis – and the quality of the services they render – are heavily dependent upon third sector handouts. For many, the 'liberalization' of the economy has further weakened an already tenuous relationship to modern state authority. We observed (but could not quantify) a migration of families, with the blessing of the Chief, away from schools and the other basic social amenities of modern civilization, into the 'deep bush' in search of fish, game and fresh forestland to slash and burn. While this may be a reasonable move from the perspective of household food security, the implied deliteracization of the population does not portend well for the consolidation of democracy nor of 'the market.'

2. For the neo-liberal development model adhered to by the MMD regime to come to fruition over the longer term, boosts in productivity spurred on by an increasingly efficient market must generate growing local incomes. Rising incomes are necessary as a basis for public taxation, and to establish the demand for private or semi-private services where the public sector lacks the means. If an accumulation process is to gain a foothold in the agricultural sector (as the current projections tend to assume), some form of specialization and differentiation must occur within the domestic economy. Chileshe Mulenga of the University of Zambia has wisely noted that in remote areas like Luapula, a low value to bulk crop like maize cannot provide the basis for commercial success. Instead, 'less bulky, high value crops such as coffee, wheat, tea and cashew nuts would have to be the main commercial crops.' Such specialization demands extensive investments in seed, fertilizer, chemicals and irrigation. Thus, as Mulenga infers, 'high value export crops would be the preserve of the better-off households. The poor households, therefore, would probably have to fall back on the better adapted traditional crops such as cassava, finger millet, mixed beans and sorghum.' (Mulenga 1993: 7).

On the basis of our data there is little evidence to indicate that the emergence of export-oriented production in Milambo, or most anywhere in rural Luapula, is a likely outcome of current trends. Despite the persistence of a degree of social differentiation, none of the 'better-off' households are in a position to invest in full-blown commercial agriculture. For the time being however, groundnuts, millet and cassava demonstrate a reasonable potential for (modest) commercial success in the regional markets of Mansa, Samfya and the Copperbelt. What can be done to improve market access and in this way pave the way for a virtuous spiral of incomes and investment? Is there any way to avoid the very real threat that the potentially lucrative trade in groundnuts, millet and cassava will not lead to rising incomes, but to a situation in which local actors with commercial (as against farming) interests ally themselves with urban-based traders to intensify crop extraction at exploitative prices?

The first consideration should be to ensure that 'the market' is itself accessible. Where the 'Humanist' regime of UNIP was chastised for its market-distorting policies, MMD's economic reforms have not fully eliminated the distortions. These may no longer derive from direct state intervention. But inadequate infrastructure, inadequate credit, and inadequate information will corrupt the market as much as do producer subsidies. Another vital issue concerns ensuring fair business practices. The co-operative marketing system of the eighties may have been rife with theft and extortion, but it did at least provide a platform for producers at the local level to share experiences and coordinate their efforts.[26] Under the current marketing regime, producers invariably find themselves dealing with far more knowledgeable, wealthier and better connected middlemen in isolation and on an individual basis.

3. Clearly, those with capital assets, and especially with a transport vehicle, are best positioned to benefit from the new market opportunities. Expenses (fuel, parts) are high, as are the risks of petty commerce, but the popularity of trading as a main endeavour or as an agricultural sideline is immense. The greatest opportunities for profit are in the hands of the urban-based trader/transporters who can choose the time, site and conditions of their business. Traders who can afford to invest in shoes, fitenge and radio cassettes in June and July when the harvest is just in, and have the capacity to store them until December or January when the price peaks, can enjoy extremely lucrative terms of trade for Milambo groundnuts. One assumes that this sort of trade is even more profitable for those who finance the middlemen. In this respect, market liberalization has no doubt favoured the 'haves,' and thus increased social polarization.

Our data suggest that local farmers have not waited idly for an external authority to dictate their economic strategies. This study sketches an image of men and women who have adjusted to the new market conditions with creativity and speed. In light of this, one could plausibly argue that the money spent on the social engineering of 'empowerment' and 'participation' could perhaps be better spent on improving the basic economic infrastructure, and improving the quality and availability of public goods. Still, the development of the local organizational infrastructure remains an important issue. The experiences of the co-operative era were contradictory, and they have not been properly assessed by local actors. To the outside observer it appears patently obvious that local communities, like individual households, can benefit more from existing marketing opportunities, and effect better deals with credit coordinators, traders and middlemen if their members collaborate rather than compete. This seems to be a lesson very hard

---

[26] I have discussed the dynamics of the local Milambo co-operative in Gould (1997).

in the learning.

4. Failure to come out with a uniform condemnation of economic reforms and SAP-style market liberalization should not be taken as an endorsement of rural development policy under the Third Republic. Indeed, it would be difficult to endorse something that hardly exists. The MMD regime, like its predecessor, has made sporadic efforts at supporting the 'modern agricultural sector,' but small-holders have constituted a residual element in these efforts. The main motive informing the current government's rural development policy, as Van de Walle and Chiwele (1994) suggest, has been to break up the clientelist structures revolving around co-operative marketing bodies built up by UNIP. Such negativism is an inappropriate and inadequate platform for dealing with the problems facing rural small-holders. One can hope that the lack of a rural policy is in part the outcome of inadequate information on the real impact of policies already put into effect. Hopefully this study will stimulate both discussion and further inquires into the situation of small-holder farmers in rural Zambia.

Annex 1: The Milambo Area

Soils, Vegetation, Rainfall
Milambo's area lies on a plateau at an altitude around 1,250 meters above sea level. Soils, while varying in specific type and texture from one location to another, are uniformly poor, heavily leached and acidic, and largely devoid of organic nutrients. The Land Resources Study of 1975 classified the entire area as exhibiting 40-59% class IV soils, which are considered to indicate 'very severe limitations for open field arable farming with or without mechanization' (Land Resources Study 1975: Map 3-2). Small pockets of better soil do occur, and local cultivators consider the sandy loams of the Totolo-Mulumbi and Masamba (Kolala) areas to be highly suitable for legumes.

Rainfall is heavy during the wet period between November and April, averaging in excess of 1,100mm per annum (Hutchinson 1972: 24). From the agricultural point of view, the timing and regularity of the rain is more important than absolute amounts, especially during the crucial planting season at the beginning of the rains. Precipitation begins in November at the latest, but the number of days in November receiving rainfall can fluctuate wildly. According to Kay there can be anywhere between 5 and 19 rainy days in an average November. An early period of heavy rain in October or early November, followed by a two- or three-week drought is also not unusual.

Natural vegetation has been transfigured radically by human livelihood. Over the last three hundred years the traditional citemene mode of shifting cultivation has eliminated all vestiges of primeval forest, leaving a 'variable Brachystegia-Isoberlinia paniculata' complex, considered to be a fire climax woodland (Kay 1964: 11; Trapnell 1953). Growing population density and increasingly concentrated patterns of roadside settlement are further aggravating this environmental trend. Within roadside corridors extending a few kilometres on either side of the main roads even the sparsely-treed Brachsytegia woodland is failing to regenerate due to intensified land exploitation and shorter fallow cycles. Woodland is being replaced in such areas by low shrubby vegetation, punctuated here and there by low canopied, fire-hardy trees and bushes.

Wildlife, once plentiful, has similarly lost in its competition with the human population. Fish is still plentiful in the Mulungushi, at least for subsistence needs, but the Lwela has become unreliable. While the availability of animal protein has apparently decreased dramatically, some indigenous foodstuffs and medicinal materials continue to be gathered from the natural environment. These include above all mushrooms, a myriad of roots and leaves, caterpillars and similar edible insects, as well as honey.

Landscape and Infrastructure

Chief Milambo's subjects have settled on and cultivated land in the Lwela River valley and its major tributaries. The landscape of this high rainfall region has been sculpted by drainage patterns as the run-off gathers into streams and rivers running along a rectilinear, east-west axis into the Lwela. Viewed along a north-south axis, the profile of the land is mildly undulating, with higher places marking the watersheds between the numerous tributaries of the Lwela. Characteristic of the Luapula region are spongy peat swamps known locally as dambos, where water collects in perennially damp, poorly draining basins along the drainage lines.[27]

The whole of the chief's area, which encompasses a large Game Manage-

---

[27] Dambos are formidable obstacles to road builders and travellers alike, and the dambo which the Milambo road traverses at the Bwingi River just north of the Chief's palace prevented the transportation of Kolala maize by Luapula Co-operative Union (LCU) lorries during the 1988 marketing season. See George Kay's thorough discussion of the role of dambos in Aushi environment (1964: 8–10). On their agricultural use see Kokwe (1991).

ment Area and extends beyond its southern boundary to Kapalala on the Luapula River, comprises 4,080 km². The focus of my fieldwork was limited roughly to that area serviced by the Milambo Multipurpose Co-operative Society (MMCS) during its heyday in the late 1980s, a surface of about 1,140 km².[28] The main concentration of the population of ten to twelve thousand inhabitants lies on either side of a gravel road which follows the western bank of the Lwela River through the five zones of Totolo-Kalonge, Mapula, Lwela, Milambo and Kolala.[29] This thoroughfare was originally built to connect the Aushi Native Authority headquarters at Milambo with the District capital at Mansa, or Fort Rosebery as it was known in colonial times. This road, which follows the western bank of the Lwela in this area, is still passable to motor vehicles in most weather as far as the Chief's palace. South of the palace (Milambo's village), the road becomes narrow and risky for motor transport, especially when crossing larger streams. Heavily traversed by pedestrians and bicycles, its terminus is at Chembe on the Luapula River.

Mulumbi zone is also accessible by the all-weather road (the D100) from Mansa that spawns the above-mentioned road to Milambo at Moffat Mulubwa about 15 km north of the MMCS headquarters in Lwela. From Mulumbi, the D100 continues east, eventually reaching the paved Samfya-Serenje highway. East of Mulumbi, the D100 crosses the Lwela over

[28] This Annex describes the Milambo area as it was at the time of my fieldwork visits in 1988–89. Changes in the administrative setup have accompanied the shift in political regime after October 1991.

[29] The most recent head count is from 1988. The total population of the area at that time was 10,830 or about 12 persons per square kilometre.

Zone	Population	% according to health census	% in survey samples
Kolala	1806	25	15
Lwela	2442	34	35
Mapula	468	7	10
Kalonge	504	7	10
Mulumbi	1488	20	24
Ng'omba	456	7	4
Total	7164	100	100

Figure 22. Structure of study sample by Milambo Primary Cooperative Society zones.

a sturdy bridge. Soon after, a neglected but motorable road branches south, through a waterless and consequently uninhabited stretch of dense forest reaching Ng'omba on the Mulungushi River after some 15 km.

Annex 2: Sample Design and Survey Methodology

Data were collected via a standardized questionnaire using trained enumerators of the provincial Central Statistical Office. The original questionnaire was in English, but as part of the pre-field training, the enumerators carried out a collective translation in to c'Aushi, which was the interviewing medium. In both instances, interviewing took place during the immediate post-harvest season, when production and marketing information was fresh in the respondents' minds and labour demands were minimal. In 1996, I worked alongside the enumerators in the field and checked their interviews on a daily basis. In 1988, I was no longer in the country when the interviewing took place, and this task was carried out by a Senior Enumerator of the Luapula CSO.

The sample was worked out on the basis of a local census carried out by in 1988, by the Department of Health. Individual villages were classified with respect to their size, composition, proximity to a marketing outlet. A set of villages was selected within each class. Within each sample village, enumerators carried out a house count on the basis of which they sampled every tenth domain. Within each household, enumerators were instructed to interview 'any knowledgeable adult household member,' preferably the household head or the head's spouse. Given the important role of women in rural agriculture, care was taken to include as many women in the sample as possible. This was more successful in 1996, when 48% of the respondents were women. In 1988, men comprised nearly three quarters of the sample. In many instances, couples (or larger groups of resident adults) were interviewed together.

The 1988 interviews were done in the wake of an intensive three-month period of informal interviews, participatory observation and open-ended discussions in the survey area by the author from May through July, during which time the questionnaire was also being designed and tested.

Sample

The sample frame was designed on the basis of an informal population census carried out in April 1988 by Fortunate Mwewa, the local Health

Assistant. Mwewa's census covered the catchment area of Lwela Rural Hospital which corresponds quite closely with the region serviced by Milambo Multi-Purpose Co-operative Society (MMCS). The health census found the total population of the area to be 9,465 persons. Of these, 7,164 were considered to fall within the Society catchment area. To facilitate sampling, the Society area was divided up into 'zones' as follows.

Households in each zone were to be chosen in like proportion to the planned total sample of 220 cases. Kolala (the southernmost zone) is underrepresented in the final sample, to the benefit of the smaller neighbourhoods of Mapula and Kalonge, as well as of Mulumbi. Ng'omba's representation also falls short of optimal. This implies that the sample may inadvertently disadvantage some of the more remote parts of the community.

Selection of interviewees was effected in collaboration with the headman in each sampled village. In 1988, the sample was also intentionally skewed to include a fifty percent representation of co-operative society members. Thus co-operative society members constitute roughly half of the final sample whereas members of only about 44% of the area's 1,325 households belong to the co-operative Society.[30] In 1996, the same sample design was employed, with the exception that the 1996 sample was no longer skewed to incorporate even numbers of co-operative and non-cooperative members.

If we assume that the Society catchment area comprises approximately 1,325 households, the effected sample represents a 17% coverage of the population. This is probably an overly optimistic estimate. The problematic nature of the concept of 'household' makes it very difficult for such generalizations to be of much practical benefit. On the one hand, a household will often comprise smaller units which may, in many important respects, be quite independent. Secondly, the pervasive gender dualism in Aushi society counsels great caution when assuming that male household members represent the views and interests of their female spouses (and vice versa). Nonetheless, if these conditions are kept in mind, and the data is approached critically it can be considered a reasonable (if not strictly representative) cross-section of Milambo Aushi population.

---

[30] According to survey data an average Milambo household has 5.4 members.

5

CHIEFS, POLITICS AND STATE FORMATION IN THE THIRD REPUBLIC[1]

(2006)

...it is easier and more satisfying to expose the contests over legitimacy than to chart the perilous course of democratization. (Wright, n.d.: 37)

The Return of the Chiefs

Kalonga Gawa Undi (b. 1931), recognized as the paramount chief of the Chewa people in Zambia passed away on 21 November 2004. The Chewa are one of Zambia's major socio-linguistic clusters and Chief Undi had an illustrious career; he was part of the delegation that negotiated Zambia's independence from Britain and he later chaired the House of Chiefs from 1968-81. Despite this grand historical stature, political commentators expressed surprise when the Zambian government organized a state funeral for the late Chewa leader – something that had not been done for any other Zambian traditional leader since independence (Simutanyi 2005). At this funeral, Zambian Vice-President Lupando Mwape proclaimed that 'Kalonga Gawa Undi X has gone at a time when Zambia requires his superior administrative skills to steer past the many social and economic pitfalls dogging the country' (Kapambwe 2004). One must be cautious not to read too much significance into a graveside eulogy. Yet, to the observer of long-term trends in Zambian politics, the imagery of local chief as national steersman signals a radical departure in official rhetoric about traditional leadership.

As in many African states, Zambia's traditional leaders were excluded from the redistribution of power after independence in 1964. Ethnographers with a grassroots perspective may attest to the 'great importance still

---

[1] This chapter is dedicated to the memories of Ndimanye Lwenje and Björn Lidgren.

to be attached to chiefs' (Van Binsbergen 1987), yet for all intents and purposes founding president Kenneth Kaunda and his United National Independence Party (UNIP) sidelined traditional leaders from the new local administrative and political structures put in place after independence. The UNIP government dealt with traditional leadership, to follow Mamdani's distinction, as a 'radical' post-colonial regime (1996: 25). The Zambian state under UNIP sought to 'reorganize decentralized power so as to unify the 'nation' through a reform that tended to centralization.' In doing so, the main institutions of indirect rule, the Native Authorities, were dismantled and their functions of control and mobilization were transferred to the grassroots party organization. Such manoeuvres reflect a general suspicion of 'ethnic' identities on the part of nationalist movements. UNIP did not go as far as TANU in Tanzania which abolished the chieftaincy, or FRELIMO in Mozambique who propagated the slogan 'Kill the tribe and build the nation.'[2] but the thrust of UNIP's political sensibility was the same. Partly retribution for their complicity in colonial Indirect Rule (Chanock 1985, Mamdani 1996), partly an expression of the postcolonial liberal state project, and partly motivated by UNIP's hegemonic aspirations, chiefs were cut off from the political apparatus of government and languished throughout the First and Second Republics in ramshackle rural 'palaces', with minimal state patronage and little real power.[3]

Unlike many other African countries, the relationship between chiefs and the Zambian state is not encumbered by the violent politicization of ethnicity (cf. Lindgren 2002 on neighbouring Zimbabwe). Chiefs in Zambia do not challenge the legitimacy of state rule, and there is no apparent threat of ethnic secessionism among any of Zambia's seventy-odd socio-linguistic groups. After the neo-liberal Movement for Multiparty Democracy (MMD) supplanted Kaunda and UNIP in 1991, however, a number of chiefs have become visibly assertive in public debates concerning issues that transcend the boundaries of their immediate constituencies, a development without precedent in Zambian politics. The MMD government, for its part, has effected a number of measures to provide greater official recognition of the chieftaincy as a public institution. In its efforts to distinguish its

---

[2] Cited by Thandika Mkandawire in a lecture on 'Decentralisation and Local Governance,' Siikaranta, Finland, May 27 1997.

[3] The exception that proves the rule was the inclusion, in 1983, of the Lozi King Lewanika and the Bemba paramount Chitimukulu in UNIP's Central Committee with the express aim, according to the party/state-owned *Times of Zambia*, of bringing the chiefs 'in the main political stream, to turn them into nationalists rather than traditionalists' (*Times of Zambia*, 29 August 1983, cited in Van Binsbergen 1987).

new 'liberalized' agenda from the authoritarian politics of UNIP, MMD promised a fresh approach to traditional leadership. The MMD campaign manifesto of 1991 announced that 'The MMD is committed to a policy whereby traditional rulers regain the enjoyment of their traditional powers. Accordingly the institution of chieftaincy shall be given its rightful and respectable role, drawing support from government' (Chanda n.d.: 69). Significantly, the 1996 constitution introduced three articles (§127-29) codifying the legal personhood of chiefs, while seven additional articles (§130-36) provide for the revival of a national House of Chiefs, comprising 27 traditional leaders (three from each province) as an advisory body to government.[4]

Chiefs have not been aloof to this sea change, and have actively sought ways to exploit new opportunities. The past several years have seen the revitalization of a national lobby organization, the Royal Foundation of Zambia, with regional bodies in a number of provinces. Through the Royal Foundation, chiefs have also begun to coordinate their political manoeuvres. When, in early 2003, Zambian President Levy Mwanawasa instituted a Constitutional Review Commission (CRC) in the face of staunch opposition by civic activists, the Royal Foundation leadership became active in brokering negotiations between dissident groups and the government.[5] The talks broke down, but four traditional leaders accepted seats on the CRC. According to the government line, Constitutional Review Commissioners were appointed in their personal capacity, but the Royal Foundation had a different take on the situation. According to Royal Foundation spokesperson Senior Chief Mukuni, '[T]he traditional leaders appointed by President Mwanawasa would effectively represent chiefs during the constitutional review process.' Chiefs, maintained Mukuni, would use their position to 'fight for the strengthening of the House of Chiefs' (Phiri 2003). This emphasis on collective political action is also a novel feature in chiefly politics in Zambia.

Against this backdrop, this essay examines changes in the role of chief-

---

[4] The draft constitution circulated by the Mung'omba Constitutional Review Commission in June 2005 (GRZ 2005) seeks to further broaden the powers of the House of Chiefs; see www.crc.org.zm (consulted 29 June 2005).

[5] The civic activists in question were an alliance of Zambia's major faith organizations, the women's movement and the Law Association known as the Oasis Forum (more on this below). The Oasis Forum was not opposed to constitutional reform, but objected to the establishment of a Constitutional Review Commission, based on the Inquiries Act, which reserved unlimited discretionary powers for the executive over the text of the revised constitution. Instead of a CRC, the Oasis Forum demanded that the task of drafting the constitution be given to a South African style constituent assembly. See (Gould 2006).

taincy, and in its impact on postcolonial state formation in Zambia's Third Republic, that is, since the restoration of political pluralism and the rise to power of the Movement for Multiparty Democracy (MMD) in 1991.[6] The historical narrative focuses on unique incidents in the local politics of Aushi chieftaincy in Luapula Province. Through this historically and culturally contextualized account, I hope to flesh out the roots of the current wave of chiefly activism in Zambia, and to point to structural continuities linking the current situation to the colonial experience and beyond.

It would seem now that Zambia's traditional leaders are getting organized, they are pooling their resources behind common political goals and their role as national political actors is being increasingly acknowledged by government. These are new and unprecedented trends in the history of independent Zambia. By calling attention to the unfolding of these processes over a longer duration, the present essay seeks to provide food for thought about the meaning of these current changes for the nature of postcolonial Zambian politics – for the constitution of power, for ideas of legitimacy, and for political relations. If and when 'traditional authority' holds increasing sway in the exercise of government, what are the implications for the trajectory of the liberal state project in Zambia which, according to ruling party rhetoric, is at the head of the political agenda? What impact, in other words, will growing assertiveness among traditional leaders have on state formation? The way one responds to these questions will largely depend on the conceptual framework through which the evidence is viewed. The following section outlines one possible perspective.

Deregulation, Retraditionalization and State Formation

> The enormous extent of [state] power cannot be understood unless state forms are understood as cultural forms, state formation as cultural revolution, and cultural images as continually and extensively state-regulated. A central dimension – we are tempted to say, the secret – of state power is the way it works within us (Corrigan & Sayers 1985: 199-200).

State Formation

Zambia's Third Republican cultural revolution is dramatic, albeit very selective. For the larger part of the rural population, 'development' is

---

[6] Zambia's Third Republic was inaugurated in 1991 with the restoration of a multi-party electoral system, banned in the Second Republic constitution of 1972.

discussed in the past tense. UNIP's ambitious rural development policies imploded in the last years of the 1980s under the weight of costly subsidies to seed and fertilizer distributors, to commodity producers, to grain transporters, to milling companies, and to urban consumers (Mickels 1997). MMD lacked the means to establish a strong rural presence and, as far as rural livelihoods are concerned, has pinned its hopes on a tiny community of semi-commercial farmers. Current reality for most rural residents is of diminished returns and evaporated amenities.

A small, if politically significant group of urban residents, however, is experiencing an unprecedented transformation of their livelihoods and lifestyles. Most of these well-educated, middle-class professionals live in Lusaka, which allows them to maintain a strong link with the forex economy – the basic precondition for membership in the new middle class. Some survive on the new business opportunities brought by new information technologies and the liberalized economy; others exploit the US$640 million per annum aid sector; yet others, like accountants, lawyers and doctors, share indirectly in the fortunes of the nouveau riche.[7] Few Zambians who live off legitimate employment are genuinely rich by G8 standards. Some members of the new middle classes, however, have achieved a standard of consumption unimaginable by the vast majority of the population. Most of them work far too hard to really enjoy their good fortunes, but their children can be seen, sporting the latest fashions, hanging at the South African-style malls, internet cafés and fast-food outlets that have cropped up in Lusaka's suburbs.

The new middle class also dominates the breaking wave of non-partisan political activism. Disenchanted with the corruption and hypocrisy of the MMD, middle-class professionals have spearheaded persistent challenges to ruling party authority (more about this in a moment). Despite strong dissident overtones, though, there is no doubt that the new middle class is the product of deregulationist policies pursued by successive MMD administrations since 1991. The initial deregulation of political life, enshrined in the 1991 constitution and expressed in the rise and fall of dozens of political parties and hundreds of new civic associations, was paralleled by an equally dramatic liberalization of the economy. During MMD's first two terms of office under President Frederick Chiluba, the government privatized much of its vast parastatal holdings, liberalized currency trading, removed trade barriers and opened the economy to foreign investment and speculative

---

[7] Based on OECD statistics consulted 20 June 2005 at http://www,oecd.ord/dataoecd/40/27/7504863.pdf.

deals (see Rakner 2004).

The pervasive deregulation of Zambian society catalysed the transformation of ideals and expectations, and transformed the core of state-society relations. The government is no longer a reliable source of livelihood for a broad stratum of workers and functionaries. There have been massive retrenchments. Public bodies no longer provide reliable health services and a decent education; instead, the private (and, for most, unaffordable) provisioning of basic services has mushroomed. The ethos of public service in government, a core principle of the Kaunda years, has given way to a view of public office as a 'job' – which in twenty-first century Zambia implies the exchange of loyalty and obedience for material reward. Without romanticizing the spirit of the Second Republic, there is little doubt that the failure of the MMD to stop the deterioration of welfare generated widespread political alienation and cynicism. By the end of Chiluba's second term in office, in 2001, his public support had dwindled to nothing. In the pre-election agitation, during which time Chiluba sought to change the constitution to allow himself a third term in office, an unprecedented alliance of civic forces calling itself the Oasis Forum emerged spontaneously to block Chiluba's bid (see Gould 2006).

MMD's popularity was also at its nadir; Chiluba's hand-picked successor, Levy Mwanawasa managed to win the 2001 presidential election (amidst widespread calls of foul play), but took office as Zambia's first minority president with less than 30% of the popular vote.[8] Throughout Mwanawasa's incumbency, the Oasis Forum and allied civic groups have consistently challenged the President and his ministers on the budget, privatization policy, on their management of constitutional reform and the war on corruption, but above all on the nature and extent of executive powers.

It seems reasonable, then, to see political restructuring in Zambia's Third Republic as both cause and effect of a transformation of ideas and practices about authority and its legitimacy, about political accountability and morality, about the benefits of public regulation, and about citizenship, rights and identity. This 'cultural revolution' does not follow a linear path toward the greater entrenchment of liberal principles in public governance for a growing number of citizens. Rather the contrary: vast numbers of Zambians have been pushed into the margins of the liberal state formation

---

[8] The real weakness of Mwanawasa's mandate is refelcted in his share of the *elgible* vote, which was no more than 15%.

project. A great many are trapped in the 'semi-subsistence' limbo of village life, cut off from the knowledge, finances and markets that could revolutionize their productivity. Meanwhile, the main – and indeed, only – challenge to a persistently autocratic style of government comes from a socially and economically privileged, and isolated, elite. Without casting doubt on the integrity of the new wave of social activists, it is also possible to see the current, Lusaka-based cycle of political contestation as a struggle among urban, professional, middle-class elites for dwindling political spoils.

Chiefs in a Deregulated Society

This condensed sketch of the Third Republic as an era of deregulation and exclusion suggests some hypotheses about the role of chiefs in the postcolonial political configuration.[9] In the language of social movement theory, the structure of political opportunity favours chiefs. UNIP's demise after 1991 left an institutional and authority vacuum at the grassroots of Zambian society that MMD never managed to fill. The retrenchments, incurred through the privatization and dismantling of public enterprises, brought many educated, experienced men back to the village. Many returnees, who are looking for productive ways to invest their savings, are willing to forge mutually beneficial alliances with their local chief, who control access to land and other local resources, thus consolidating and extending the scope of traditional authority.

The rural base of the chieftaincy has thus consolidated itself – by default if not by design – over the course of the Third Republic. Chiefs are also making inroads into national political arenas. It does not seem implausible that the next decade will see a further consolidation of these inroads. For example, the 2005 Constitutional Review Commission – in which four influential chiefs took part – purports:

> Traditional authority, customs and practices need not be incompatible with demands of modern democracy and development. What is required is an innovative institutional arrangement which combines the natural capacities of both

---

[9] One must note the problems in talking generically about 'the chiefs' in Zambia. The 286 gazetted (state-recognized) traditional leaders comprise a very heterogeneous group - there are 'kings' and 'paramount' chiefs that rule over historically regimented bureaucracies and hierarchical social structures; and there are chiefs created a century ago by colonial decree to administer peoples with little historical experience of centralized rule. Indirect Rule both generated and codified local differences. The subsequent exclusion of traditional authorities from the nationalist project has also spared (or deprived) them of the homogenizing effects that can accrue through integration into a transterritorial infrastructure of local government. The 'royal community' is steeply stratified, and the public role of individual chiefs vary greatly: the Lozi king (Litunga), and the Bemba, Ngoni and Chewa paramounts are not active public figures. For the Litunga, it would be traditionally impossible - he is only heard through his Ngambela. Other chiefs have similar restrictions on their public appearances.

traditional and elected local government to advance the development of the communities (GRZ 2005: 557).

Many would disagree with this understanding of the relationship of traditional authority and democratic rule. The following section examines the nature and implications of this disagreement.

Retraditionalization

Public discussion of the role of chiefs in Zambian politics is largely carried out in the daily press.[10] Media debates reflect the wide public interest in chiefly affairs, but the journalistic focus limits the analytical scope of the discussion.[11] Recent experiences elsewhere in the Southern African region have had a major impact on how these issues are framed in the academic literature. Indeed, scholarship on and in the region is dominated by empirical analysis and reflections on the South African situation. Within this context, debate is largely driven by the fact that post-apartheid constitutions and subsidiary legislation in both South Africa and Namibia have afforded central legal status to 'customary law' and, consequently, to local chiefs who administer it (see e.g., D'Engelbronner-Kolff et al 1998). South African chiefs continue to enjoy rights, amenities and political powers at the local government level conferred upon them under apartheid-era legislation, and this has raised political ire, especially among the circle of radical modernist analysts. The terms of this discussion would seem to be of limited relevance in the Zambian context, where the role of chiefs in local government has yet to become a burning political issue. Still, with increased networking among traditional leaders in the Southern Africa sub-region, the South African experience has had a direct impact on the politics of chieftaincy in Zambia as well.[12]

The relationship of traditional leaders to South African local government administration has been characterized as a problem of 'two bulls in one kraal' (Oomen 2000). This approach posits an intrinsic tension between customary authority on the one hand and the institutions of representative

---

[10] One of the few publications considering the relationship of chiefs to politics in the Third Republic is the proceedings of a 1997 workshop sponsored by the conservative German Konrad Adenauer Foundation, *Traditional Leadership and Democracy in Zambia*.

[11] The Mung'omba Constitutional Review Commission (GRZ 2005: 549) received 1,229 submissions related to traditional authority between 2003 and 2005.

[12] In May 2004, Zambia's Women for Change hosted the first meeting of the SADC Council of Traditional Leaders in Livingstone under the theme of 'Promoting good governance through traditional leaders.' Several dozen chiefs from all over the Southern African Development Community, from South Africa to Angola, attended the meeting which focused on sharing experiences concerning chiefs' relationship to their respective governments.

democracy on the other (Ntsebeza 1999). As conceived by Western liberal political theory, the institution of chief, a hereditary position grounded in custom and ascribed properties with strong spiritualist overtones, is anathema to the basic principle of political legitimacy in an electoral system. On this basis, the radical modernist position argues that residues of customary rule constitute obstacles to liberal government and must be exorcised from African politics (Mamdani 1996, Munro 2001; see also Etounga-Manguelle 2000). An opposing view argues that only through the restoration of the fundamental values expressed through community consensus can Africa achieve true democracy (e.g., Coalition of Traditional Leaders, 2002: 4; Chifuwe 2003). This definition of chiefly rule as 'direct democracy' implies a privileged political role for traditional leaders.

A number of writers have looked for a way around this normative impasse. One tact has been to defuse the juxtaposition of 'tradition' and 'democracy' by questioning the traditionalism of the new breed of traditional leaders. Building on the seminal analyses of colonial and postcolonial 'invention of tradition' (Hobsbawm & Ranger 1983), a number of scholars argue that contemporary chiefs should not be equated with the traditional leadership of the primordial past, nor even of the 'traditionalized' colonial/apartheid dispensation, but should be seen to represent a reconstituted, i.e., 'retraditionalized' form of custom that is largely modern in its ideological conception and political deployment.

In her case study of one South African chieftaincy, for example, Oomen underscores the fluidity of tradition: chiefs are seen to be embedded in their communities, and the scope and substance of their authority is subject to perpetual negotiations among national state bodies, local government authorities, civic leaders and a plurality of ethnic and political entrepreneurs. Far from being a fixed, codified domain overseen by despotic authority, custom is the subject of intense social construction: 'Not only is there a debate on what the real tradition is... but also on what it *should* be' (Oomen 2000: 90). Oomen's analysis suggests that the pluralist framework of the South African post-apartheid state project circumscribes the agency of traditional authority within a grid of political liberalism. In order to maintain (or preferably expand) their current constitutional mandate, chiefs relate to the national state and other local civic structures within rules and procedures defined by the state's liberal ideology.

According to Koelble some traditional leaders are indeed 'attempting to 'democratise' their role and function in rural society' (2005: 4). At the same time, however, 'others wish to establish "commandement" [invoking Mbembe 2001] structures of authority and power that replicate colonial and apartheid experiences.' Debates about the politics of chieftaincy tend to oversimplify reality since much of the discussion ignores 'the complex and hybrid nature of traditional leadership, its enormous variations across the region, its adaptability and malleability' (p. 4). Koelble is critical of commentators (e.g., Munro 2001) who make generic claims about the incompatibility of traditional leadership with the liberal politics of South Africa's current leadership. Chiefs, claims Koelble, have strong support in South Africa's ruling party, the ANC, and traditional leadership has a firm basis in the 1994 Constitution and in the subsequent Traditional Leadership and Governance Framework Amendment Act of 2003. Juxtaposing the 'illiberal' politics of chieftaincy against the liberal state-building project of the ANC, then, misrepresents the political dynamic of the emerging order in which the distinction between forces of progress and reaction does not follow lines familiar from western contexts. Koelble's noteworthy point is that defining chieftaincy and democracy as incompatible 'presupposes a linear progressivity from dictatorship to democracy, from tradition to modernity. There is no room in this model for the actual condition of the post-colony, namely hybridity' (p. 20).

Hybridity, of course, can be (and is) invoked to muster an overblown relativism which turns a blind eye to persistent residues of the 'separate but equal' treatment of different classes of citizens. This would not seem to be Koelble's intent; he has little sympathy for chiefly claims to legitimacy based on their embodiment of 'direct democracy.' He acknowledges, however, that 'democracy in South Africa is bound to take a different course from those in Europe or North America' (p. 24). Rather than allow oneself to be embroiled in ideological debates about what constitutes 'genuine' democracy, Koelble eschews such posturing and argues instead (echoing historian Marcia Wright's epigram above) for the importance of understanding 'how different political actors use the term 'democracy' to engage in a battle for political power and authority' (p. 3).

Clearly, conceptualizing postcolonial African state formation as rooted in liberal principles while also accommodating a residual domain of retraditionalized chiefly authority begs the question of what liberalism implies in this historical context. No one, from rural chiefs to the highest echelons of the ANC, is likely to have a definitive answer to this question; everyone is too busy working out the modalities of their various political identities and strategies in a constantly shifting environment. Furthermore, these complex processes of analysis, strategization and self-invention are themselves central to emerging forms and modalities of stateness. In any event, the politics of liberalism is certainly more complicated than imagined in conventional 'rights-driven' approaches to depoliticized 'governance.' Indeed, the chieftaincy dilemma cannot be resolved – in theory or in practice – if one overlooks the apparent interpenetration of 'liberal' and 'illiberal' modes of political agency.

Zambian Exceptionalism

The specifics of the political context must be taken seriously. South African debates are driven by the notoriously heated politics of authority and accountability in local government. By the same token, local government is a key focal point for contestation about the new – and inevitably selective – politics of empowerment and inclusion in post-apartheid South Africa.

Zambia's case is decidedly different. Devolution of powers to District Councils has never generated much enthusiasm among Zambia's political leadership. The ruling party controls discretional patronage through Members of Parliament (e.g., Constituency Development Funds), while the rural investments of foreign donor agencies are largely channelled through central government ministries or NGOs. As a result there is no real competition between elective and royal authority at the local government level – neither chiefs nor Councils have resources at their disposal to build up and maintain a grassroots clientele.

The main political tension in play thus concerns the relationship of chiefs and the Zambian state. From the government's point of view (that is, that of the ruling MMD), chiefs represent an authoritative link to the rural grassroots – the Achilles heel of government – and thus a potential threat to the ruling party's hegemony. MMD's strategy (if any) is to integrate chiefs into the state apparatus in such a way that would preclude any overt challenges to party/state authority at the grassroots. Under Chiluba's reputedly 'kleptocratic' administration, this took rather brash forms. According to sworn testimony by Xavier Chungu, Chiluba's chief of intelligence, numerous chiefs were given 'vehicles and substantial amounts of money... before elections for them to influence their subjects to support the MMD'

(*Times of Zambia*, 31 January 2003).

The chiefs for their part (at least Royal Foundation activists) hope to exploit the government's grassroots insecurity, and use it to consolidate an institutional position from which to negotiate for greater recognition of their interests. The House of Chiefs (HoC) has been the primary convergence point for these two strategies. Under Chiluba's administration, government consolidated the HoC's statutory position, while chiefs focussed their attentions on making it a functioning advisory body.[13]

Since 2001, however, the Mwanawasa administration's wobbly political footing has opened up space for a new mode of non-partisan oppositional politics, epitomized by the Oasis Forum (as noted above). The Oasis Forum's challenge to the political establishment has centred on constitutional reform, and above all on the need for a popular constitution, based on the broad-based participation of citizens from all walks of life in its enactment. The Oasis Forum's call for a South African style Constituent Assembly is a direct challenge to the virtually limitless powers of the Executive, but it also underscores the deepening political marginalization of large tracts of the rural and urban poor, and thus cuts to the core of the political asymmetry of Zambian society. Chiefs have followed these developments closely and weighed the implications of this process for their own strategic interests, but their allegiances appear to be in conflict. On the one hand, their primary loyalties lie with the state, the main source of their patronage. This explains why the Royal Foundation let down its civic allies and accepted positions on Mwanawasa's Constitutional Review Commission. On the other hand, in acknowledging the chief's participation in the CRC, the Royal Foundation also put one foot behind the Oasis Forum's demands: 'We will call for a constituent assembly because this [Constitutional Review] commission is just a working group,' Chief Mukuni said, 'We demand for a constituent assembly' (quoted in Phiri 2003).

For the purposes of this discussion, the intricacies of actors' evolving strategies are less important then the more general pattern: The emergence of a group of 'national' chiefs who, over the past few years, have managed to achieve significant visibility in debates of importance to Lusaka-based political society. Key players in this group include Lozi Senior Chief Inyambo Yeta (who co-chairs the CRC and chairs the House of Chiefs), Ila Chief Bright Nalumamba (another CR Commissioner and Royal Foundation President), Soli Chieftainess Nkomeshya (also on the CRC and once a Minister of

---

[13] The newly constituted House of Chiefs became functional in December 2003 (GRZ 2005: 546).

State under Kaunda), and several others. Significantly, the figureheads of Zambia's main socio-linguistic groups (e.g., the Litunga, Chitimukulu, and Mpezeni) do not figure among the ascendant group of 'national' chiefs.

The national chiefs that have emerged as brokers on behalf of Zambia's heterogeneous royalty share many features. They are politically ambitious.[14] Few, perhaps, harbour personal pretension to high office, but they are keen to exploit opportunities to act as brokers, allying themselves with political actors who can support their many-stranded interests. Their mobility and visibility is based on advanced educational qualifications and, often, donor connections (education and donor patronage are virtually synonymous), primarily through NGO mediation. They thus share many of the cosmopolitan features of the professional elites of Lusaka (higher education, knowledge of English, good PR skills, understanding of the business of 'projects'). Despite their elite habitus, however, there is no indication that the national chiefs would not enjoy the overall support of Zambian royalty for their main strategic endeavours – i.e., consolidating and expanding the powers of the House of Chiefs.

Zambian chiefs are consciously and strategically crafting a novel mode of national political agency in the deregulated, pluralist political dispensation of the Third Republic. This raises interesting questions about the insinuation of 'traditional authority' into the fabric of the liberal state-building project of Zambia's cosmopolitan elites and their donor allies. Unlike their South African colleagues, Zambian chiefs are not struggling to hold onto colonial/apartheid-era privileges; these were dismantled 40 years ago. How does one explain their current visibility, assertiveness and apparent success? Transnational support has undoubtedly been an important catalyst for the emergence of chiefly political agency within the national political arena. Zambian chiefs have been much encouraged by the work of the SADC Council, and over the last decade a number of bilateral donors and international foundations have also been working through local associations to 'normalize' chief–state relations.[15]

It would be a mistake, however, to see the reconfiguration of state-chief

---

[14] Inyambo Yeta, for example, accepted the vice-presidency of UNIP in 1995 (only to be arrested and imprisoned on trumped-up treason charges in the run-up to the 1996 elections).

[15] For example, the German Friedrich Ebert and Konrad Adenauer Foundations and the Open Society Foundation have worked with, among others, Women for Change and the Zambia Independent Monitoring Team on projects promoting the role of chiefs. In the 1990s, the Norwegian government funded dialogues between chiefs, councillors and district administrators in Northern Province. Chiefs have been actively involved in countless donor/NGO HIV conscientization projects (see 'Traditional Leaders in Zambia: A Weapon against AIDS').

relations as externally driven. The ensuing narrative about a group of Aushi chiefs in Luapula Province attempts to show that national chiefly agency has been in the making for some time. It suggests that these renewed claims by chiefs to political influence find expression in grassroots institutions – that is, in the very processes through which social and political order is established. This portrayal of a groundbreaking episode in rural politics will no doubt raise more issues than it resolves. Does the embeddedness of the chieftaincy in local clan and lineage structures, and their roots in specific tracts of soil and bodies of water, ensure that chiefs remain accountable to their rural constituencies? Will their incorporation in Lusaka political society inevitably lead to the alienation of chiefs from their local constituencies? As chiefs consolidate their political agency in national affairs, will their impact be conservative, enhancing entrenched predilections for populism and patrimonialism? Or will they apply their popular mandate to the progressive reform of liberal politics, propelling political alliances committed to decent terms of agrarian trade, job creation and equitable economic development? What forms will hybridity take in the Zambian context? Can one chart a path toward greater democracy in Zambia that builds on the agency of traditional leaders?

Aushi Chiefs in the Third Republic

Chief Mabumba's Way

In 1992, the incumbent Aushi chief Mabumba V (Musa Kalmimankonde) passed away.[16] He was succeeded by his nephew, Kalaba Chenga, who thus became the sixth Mabumba. Prior to acceding to the chieftaincy as Mabumba VI in 1992, Kalaba Chenga demonstrated an interest in the opportunities of the modern political realm. This is evident from his success in securing a coveted UNIP Ward councillorship on the Copperbelt in the 1980s. A progressive mentality may have been a family trait, for his uncle Musa Kalimankonde (Mabumba V) had also been attune to the spoils of secular power. As Chief, Mabumba V was an advocate of 'development' and had actively worked to cultivate a symbiotic relationship between his offices and the state. He managed to promote himself, by virtue of his superior command

---

[16] The Aushi are a matrilineal people who inhabit the southern plateau of Zambia's Luapula Province. Aushi reside primarily in Mansa and Mibenge Districts. A distinct, but closely related group of c'Aushi speakers is also found across the Luapula river in the Congolese region of Katanga (Shaba). For more detail, see Gould 1997.

His Honour the Royal Highness Senior Chief Milambo[1]
Fellow Traditional Chiefs
Honourable Minister[2]
Your Lordship the Bishop
Government officials
Party officials
The Officer-in-Charge, Mansa Police
All Headmen
Ladies and gentlemen:

We are gathered here today on the 30th of August 1996. This is the first time that we are remembering our old traditional ceremony of celebrating the Makumba-wa-Chilololo-Chalwe-Bulimi which was being celebrated by our ancestors. Makumba-wa-Chilololo-Chalwe-Bulimi is an old traditional ceremony which was practised by old Aushi Chiefs since the time they came to Zambia from Kola.[3]

This custom was forgotten in 1932, with the death of the custodian of the custom, Chief Mabumba II (Kalaba Musenga). All these past years, Aushi Chiefs worried about the revival of this important traditional ceremony. So this ceremony will now be an annual event. In order for this ceremony to work annually, I am appealing to all my fellow Aushi Chiefs to work as a team and unite to see the success of this event. This will help our own children to remember the traditional values of the Aushi people as is the case with other tribes in our country.[4]

In order for this event to work, we are asking the government to involve itself by bringing development to our respective villages, e.g.:

1. Provide better feeder roads in Chief's areas,
2. Electrification of Chief's Palaces as is the case in the Luapula valley,[5]

---

[1] Chief Milambo denies having been present at the ceremony.

[2] This was the Provincial Deputy Minister who is the most senior MMD representative in the province. It is unlikely that the Minister was present at the ceremony.

[3] 'Kola' refers the mythical ancestral home of the Aushi people. The Aushi are thought to originate in the Lunda-Luba kingdom located in Eastern Zaïre.

[4] The Makumba-Chilololo-Chalwe-Bulimi celebration is explicitly modelled on similar 'traditional' ceremonies that were incorporated as national cultural artefacts during the UNIP era. The closest example is the Mutomboko, which is celebrated by the Luapula Lunda. According to Chinyanta and Chiwale, the Mwata Kazembe of the Lunda dances the Mutomboko to commemorate the Lunda crossing of the Luapula River upon their arrival from Kola, and their subsequent victories against their enemies. The 'modern' Mutomboko is danced on 29 July to honour the instalment of Kazembe XII (Kanyembo VI) on that day in 1961; see Chinyanta & Chiwale (1989) and Gordon (2004).

[5] This refers to the differential treatment that the southern or 'Plateau' Chiefs (Aushi, Kabende, Chisenga) feel to be bestowed on the northern or 'Valley' (Lunda) chiefs

3. Introduce markets in villages,[6]
4. A person cannot work on an empty stomach' so the Government should assist small-scale farmers with farm inputs with no strings attached to the loan. Some lending institutions just steal the farmer's money without giving them farm inputs as was the case of one lending institution (AMBER) during the 1995/96 farming season.[7]

Lastly I wish to thank all those who worked so tirelessly to prepare this event and those who gave out various gifts, and those who provided transport, etc. On this we are saying thank you for a job well done. I would also like to extend a hand of thanks to the 'Makumba organising committee' for all the work that they did to make this event a success.

Thank you.

within Luapula. The Lunda Mwata (Chief) Kazembe is classified as a Paramount Chief putting the Lunda one notch above the Aushi whose leader is only a Senior Chief. Historically, the Luapula valley benefited handsomely from its rich fisheries, and for a period in the eighteenth and nineteenth centuries, Kazembe's Lunda kingdom was a fulcrum of the long-distance East-West Arab-Portuguese trade that transected Central Africa. The Luapula valley has thus been a perennial seat of economic and political power in the region. Current jealousies are further fuelled by the fact that President Chiluba has Valley roots.

[6] This is a call to reinstate the depots for agricultural produce that disappeared with the advent of market liberalization in the early 1990s.

[7] Liberalization also meant the end of subsidized seasonal loans for small-scale farmers.

Figures 22 (previous page) and 22a. Chief Mabumba's speech at the 1996 Makumba–Chilololo–Chalwe–Bulimi–Bwaushi ceremony, Chief Mabumba's Place – Mansa. With notes.

of English, as the representative of the Aushi chiefs on the Mansa District Council, over and above Semu Nkandu, then government-approved Aushi Senior Chief Milambo (Gatter 1990: 384).

By the end of the 1980s, Musa had succeeded in attracting a 'disproportionate amount of the district's development resources' from donors and donor-supported government institutions (Gatter 1990: 385-6). Musa's modernism did not imply a rejection of his traditionalist habitus. An integral part of Musa's development strategy was to restore 'Makumba' to the area of his chiefdom.[17] Philip Gatter reports that during his fieldwork in 1987/88, 'the news was spreading that the re-established relationship [between Mabumba and Makumba] was at last to be cemented by the building of a new shrine in Chief Mabumba's area, thus literally and metaphorically bringing the spirit home' (Gatter 1990: 380).

Musa's vitality ran out before he could see this dream to fruition. He did, however, convey his plans for the Makumba shrine to his nephew and successor Kalaba Chenga (Mabumba VI). Kalaba Chenga, with the political zeal of an ex-District Councillor, took up his uncle's torch. By 1996, he had rallied sufficient support to ensure government endorsement of a neo-traditional '*Makumba-wa-Chilololo-Chalwe-Bulimi*' ceremony on the Mabumba palace grounds.

Kalaba Chenga claims that Musa left papers with him explaining the Makumba ceremony and other items of Aushi culture. From these notes Kalaba learned that there was an ancient Aushi ceremony of celebrating Makumba. On assuming office as Mabumba VI, the Kalaba met in Mansa with a number of other Aushi leaders to discuss these plans.[18] The Chiefs agreed that the Makumba ceremony must be revived. They also concurred that the celebration had always taken place in Mabumba's area, and should continue to do so.

During the course of 1996, Chief Mabumba mobilized his subjects to make a large clearing behind his Palace, to build fencing around the ceremonial area and to erect thatch structures to accommodate the guests of honour. Mabumba also put together an organizing committee of well-to-do Aushi businessmen based in the provincial boma of Mansa and found suffi-

---

[17] Makumba is the foundational totem of Aushi cosmology. Makumba has been described, alternatively, as a tribal god (Philpot 1936), a divinity, a protecting spirit (Labrecque 1982), a totem (Verbeek 1990), an ecological shrine (Cunnison 1959) and a territorial cult (Van Binsbergen 1981); see Gould 1997 for greater detail.

[18] The Chiefs who met with Mabumba were Mibenge, Kale, Chimese, Kalasalukangaba and Kalaba. Together these represent about half of the senior Aushi leaders.

cient resources for publicity, transport and refreshments on the day of the ceremony. According to Chief Mabumba's press release, the neo-traditionalist Makumba ceremony featured 'traditional dances and an exhibition of traditional Aushi utensils,' and was put on for the benefit of 'tourists and all traditional Aushi chiefs' (*Daily Mail* 27 August 1996).' Chief Mabumba's own explanation of the intentions behind 'reviving' the Makumba ceremony is straightforward. In his words, 'Makumba's ceremony was intended to call Makumba back [to Mabumba] and find a secret place for it to rest.'

On the basis of the speech, one might discern some other motives as well. Mabumba's call for the Aushi Chiefs to 'work as a team and unite' is a thinly veiled appeal for the *benang'ulube* to acknowledge his leadership. Mabumba gives two reasons why stronger leadership of the Aushi people is needed. First, their subjects are suffering due to the withdrawal of State support to rural development. The roads are dilapidated, and market liberalization (with the concomitant removal of agricultural subsidies) has left the district's small-holders without a market for their surplus. Small-scale producers cannot even obtain seasonal credit for their farms. Second, Mabumba raises the sore issue of the deprivation of the Luapulan peoples living on the central and southern plateau in comparison with the situation of the Luapula river valley Lunda of *Mwata* (chief) Kazembe. The valley road was paved in the 1980s and electricity is widely accessible. The general view among the plateau peoples is that none of their own has ever been selected to serve among the province's political or administrative leadership. In contrast, numerous provincial permanent secretaries and ministers have come from the valley. The fact that then President Chiluba also originated from a Lunda village in the Luapula river valley (potentially a source of increased patronage for all Luapulans) further exacerbates the frustration felt by the plateau folk. Mabumba's message is clear: Let us unite under the auspices of our ancient heritage to promote our common goals. By claiming success in 'calling Makumba back,' Mabumba suggests that he is the genuine leader the people need to defend their interests.

This helps explain why possession of Makumba is so important. For Mabumba, as for all the royal *benang'ulube*, Makumba embodies the source of the spiritual powers for which the true leader is the prime conduit. Makumba will only reveal itself and provide counsel to someone with the supernatural powers to commune with the greatest of the ancestral spirits of the Aushi people. For all their other differences, Milambo and Mabumba agree on this principle. The question is, who has these powers and how might this be proven?

Possession of Makumba certifies the legitimacy of one's claim to the senior chieftaincy. This may be an end in itself. But for a political entrepreneur like Mabumba, the point is that the senior chieftaincy represents a rare form of political capital that appears to be rapidly appreciating under the Third Republic. Zambia's first President Kaunda's tactic was to subordinate Zambia's traditional rulers to absolute Party discipline. Under the UNIP regime, unlike today, chiefs could contest parliamentary seats and several traditional leaders doubled as members of parliament. Under the strict political discipline of the one-party state, this did not imply a significant exercise of discretionary power. On the contrary, Kaunda's approach subverted any overt challenges to his authority while conveying a powerful symbolic message to rural subjects about the subsidiarity of their local leaders. In his immediate dealings with chiefs, Kaunda was careful to grant them the traditional trappings of respect. But in terms of political or administrative authority, chiefly powers were limited to usufructary land distribution and arbitration in minor civil disputes.[19]

The tensions between the UNIP Party-State and the chiefs fermented just beneath the surface. In late 1988, for example, the UNIP Party head in Mansa District was called on the carpet for his lack of respect to Chief Milambo. A UNIP deputy minister of Luapulan origins filed the following minute to the district governor:

> When you [recently] addressed a meeting at Chief [Milambo's] palace, you told your audience that you had powers to dethrone Chiefs who did not follow Party regulations and that Senior Chief Milambo was one of them.[20]

As a result, Chief Milambo became so angry, noted the Minister, that he was 'reported not to be eating.' The Minister advised the Governor to go see the Chief for a 'private chat,' to proffer his private apology for his public display of disrespect. There was, however, not a directive for the Governor to publicly retract his threat. The Governor's statement, while indiscreet in its bluntness, conformed to UNIP's view on the relationship between Chiefs and the State.

From its beginnings in 1990, MMD sought to mine the grassroots discon-

---

[19] In most of rural Zambia, land for semi-subsistence cultivation has been abundant, and the right to allocate land for use has carried more symbolic than real political weight. This right has resided with the Chiefs, though it has generally been deployed by village headmen. Market liberalization and the new Land Act of 1996 that vests all land in the President may lead to sweeping changes in the value of land. In the short term, this is only likely to affect areas adjacent to major markets.

[20] National Archives of Zambia, Lusaka, LP/SEC/28/2, folio 160, 23 December 1988.

tent engendered by such demonstrations of arrogance. The MMD Manifesto appealed to the traditional leaders for political support and promised to restore their diminished powers.[21] This policy apparently filtered down to the operational level. Indeed, the Mansa District Secretary told me in 1996 that 'MMD is committed to empowering the chiefs. In some respects the current policy is a throwback to the colonial practice of Indirect Rule.'[22]

Indirect Rule was not, of course, an instrument of 'empowerment.' One cannot be completely sure whether statements of this sort express irony or historical naïveté. From the perspective of the traditional leader, however, the opportunities afforded by the rhetoric of retraditionalized Indirect Rule are real. Under MMD, chiefs are called to sit on committees that allocate 'Constituency Development Grants', one of the main instruments of MMD's political patronage. It is obvious to the grassroots actor that UNIP is gone, and nothing has taken its place. After 80 years of manipulation by the colonial and post colonial States, the struggle to regain tangible power is just beginning. The first move for an ambitious political entrepreneur like Kalaba Chenga is that of capturing the strategic pivot of the Senior Chieftaincy. In this, he chose Makumba as the key symbol of his campaign.

Milambo's Rebuttal

The main obstacle to Kalaba Chenga Mabumba's political ambitions was the incumbent Senior Chief Milambo. The incumbent Chief Milambo at the time, Darius Kapesha, was in his 70s when he succeeded to the chieftainship in 1994. By most accounts, Kapesha should have become Milambo in 1953, when the incumbent (Cilyapa) was banished by the colonial authorities for his opposition to Federation.[23] Kapesha was passed over by the colonial administration after Cilyapa's banishment by virtue of his political connections with the nationalist ANC. Several *benang'ulube* elders claim that Cilyapa deposited Makumba with Kapesha for safe keeping upon his deportation to Luwingu. Indeed, Kapesha was widely considered to be the legitimate heir to the chieftaincy, especially after the death of Cilyapa in 1977. That this is so is substantiated by the fact that Kapesha is known to have presided

---

[21] MMD's policy on the Chiefs is an isolated enclave of sensitivity to rural issues amidst a decidedly urban agenda. It is somewhat ironic that the section in the Manifesto on the Chiefs was drafted by Guy Scott, a white Zambian commercial farmer and MMD's first Minister of Agriculture. Personal communication from Guy Scott, 8 April 1998.

[22] Interview with Mansa District Council Secretary, September 1996.

[23] The Federation of Rhodesia and Nyasaland lasted from 1953–1963. It was widely unpopular amongst the African s in Northern Rhodesia (Zambia) and Nyasaland (Malawi) for fear it would lead to full amalgamation with Southern Rhodesia and white minority rule.

over key Aushi rituals – such as the restoration of the symbolically powerful *babenye* shrine – while Semu Nkandu was still alive.

Kapesha did not watch Chief Mabumba's manoeuvres disinterestedly. When asked to comment on Chief Mabumba's Makumba campaign, his assessment was blunt: 'Mabumba organized this ceremony at his palace to drum up support because he wants to grab the Senior Chieftainship from me.' Chief Milambo rejected the validity of Mabumba's claims in no uncertain terms. In his words, 'Chief Mabumba has never kept Makumba as the current chief claims. Makumba has been here at Milambo all this time.'

The ultimate powers of arbitration in a dispute over the Senior Chieftaincy would reside with the *benang'ulube*, the main power brokers of which are the incumbent chiefs. The politics of the struggle are clear in the Senior Chief's mind: 'Chiefs Kalaba, Kalasa Lukangaba, Mibenge, and Chimese attended the (Makumba) ceremony. Chiefs Matanda, Kasoma Lwela, Sokontwe, Kundamfumu and Kafwanka did not' (and are thus Kapesha's allies). Senior Chief Milambo himself boycotted the ceremony 'because the ceremony wasn't supposed to be held there since Makumba has never been kept at Mabumba.'

Historical knowledge – or better, interpretation – constitutes the main weapons employed in such a political struggle.[24] Milambo and Mabumba presented carefully rehearsed versions of the critical juncture in the saga of Makumba: everything hinges on what happened to Makumba when Kalaba Musenga (Mabumba II) died in 1932. Here is Mabumba's rendition:

> After the death of Kalaba Musenga, another Chief came to take the throne. He didn't follow proper procedures. At that time Makumba was celebrated here [in Mabumba], not at Milambo. Makumba became annoyed at the new chief.
>
> After that some people thought that Makumba should be taken to Milambo. Milambo came to fetch Makumba by bicycle, but Makumba ran away and came back here to Mabumba. Makumba could not be carried on a bicycle nor wrapped in a cloth. He can only be carried 'on the shoulder.'
>
> Makumba was not happy where it was being taken by [Milambo]. It was changing to different forms indicating its dissatisfaction. The kidnappers said, why is it changing, and Milambo tried to crush it, but failed. Makumba escaped and came to rest at ... a secret place that someone like you could never find, that is not burned.[25] Makumba came to rest there. Since then it has been neglected and it has been moving up and down but within Mabumba's area.

---

[24] This insight is now, of course, a commonplace in political anthropology. Although note that the theoretical foundations of this view of the efficacy of historical consciousness in local politics were laid by Ian Cunnison's (1959) work on the Luapula Lunda.

[25] 'A place that is not burned' is one untouched by *citemene* (swidden) cultivation, that is, far from (or magically invisible to) human habitation.

Predictably, Chief Milambo has a competing view:

> Chief Milambo Chilyapa was the custodian of Makumba Chilololo Chalwe Buli-mi. It used to be kept in the palace grounds here in Milambo ... Chief Mabumba has never kept Makumba as the current chief claims. Makumba has been here at Milambo all this time.[26]

Indeed, Chief Milambo disputes the whole idea of the Makumba-Chilololo-Chalwe-Bulimi ceremony. According to Milambo, in previous times 'there was no celebration of Makumba. There was just a thing called *ichimuno* whereby all Aushi chiefs and their councillors could assemble' to consult with the *chilaluka* (a priest or oracle) of Makumba. Indeed, while Verbeek mentions seed blessing and first harvest ceremonies centred on Makumba, there is no mention of any such public spectacle in the existing literature on Aushi traditions.

Soon after becoming Mabumba in 1992, Kalaba Chenga claims to have seen Makumba by a stream near the Mabumba Basic School which is adjacent to his palace grounds. There he encountered Makumba

> ...in the form of a snake, spitting saliva. I was clapping my hands saying 'Yes, I know you.' Makumba started making a sound like an engine. I understood this as my being welcomed by Makumba as Chief.[27]

Having thus established the pedigree of his claims to Senior Chieftainship, Mabumba was prepared to provide the historical justification for Makumba's blessings:

> In colonial times, everything was prepared for Mabumba to be made Senior Chief, but he was late for the meeting, so the District Commissioner gave the inkondo [a small stick] to Milambo since he was the only one to attend. The people were annoyed. Even now the Chiefs know that Milambo was only appointed by the colonial government. Mabumba is the real Senior Chief.[28]

Chief Mabumba had a clear retribution strategy already worked out. His plan in 1996 was to call together 'all bwaushi chiefs to go back to the beginning to see how the senior chieftaincy was determined.'[29] He seemed confident that from his current position of strength, he could reverse the injustice dealt to his office by the colonial district commissioner nearly a century ago.

Political chiefs

In 1996, Chief Milambo also had plans to call up a meeting of all Aushi chiefs,

---

[26] Chief Milambo, 22 September 1996.

[27] Clapping hands (while kneeling) is the proper form of respectful greeting in the presence of an Aushi Chief.

[28] Chief Mabumba, 26 September 1996.

[29] Ibid.

headmen and chiefs' councillors. The point of his conference would not be to discuss the senior chieftaincy, but to take disciplinary action against Chief Mabumba for his unbecoming behaviour. This was to be after the presidential elections which were held in November of that year. When the Aushi chiefs did finally come together after the elections, Milambo's disciplinary action did not appear on the agenda. Instead the chiefs found themselves entering a new historical era of political intervention. The purpose of the meeting, in which Mabumba appears to have been the leading player, was to rally the chiefs behind an assault on the provincial MMD leadership.

Kalaba had problem with the deputy minister, the top MMD official in the province. There was a botched coup attempt by disgruntled junior military officers on 28 October 1997. After the event, someone spread the rumour at the boma that Mabumba (known to be a former UNIP politician) had openly rejoiced and 'clapped his hands' upon hearing the news of the coup.[30] As a result, the deputy minister began to level treason accusations at Chief Mabumba and to harass him and his advisors.

Mabumba was not the only one in the bad books with the minister. The permanent secretary (the head of the provincial civil service) is also embroiled in a drawn out feud with political counterpart. The source of the hostilities is too manifold to contemplate in this connection. As far as I can fathom, the P/S's dislike of the minister derives from the fact that the minister comes from the same village as, and is a classificatory 'uncle' to, President Chiluba. This is not, generally speaking, a liability in Luapula. That the minister routinely evokes this connection to 'pull rank' on his compatriots has nonetheless made him universally unpopular. With the permanent secretary, the friction purportedly exploded into a tilt of public insults at the Mwata Kazembe's Mutomboko ceremony.

Some conniving between the permanent secretary and Chief Mabumba was possibly involved when ten Aushi chiefs came together in February 1998 to draft a letter to the president.[31] In the letter (see Figure 23) they lodge a stringent complaint about the deputy minister's treatment of the chiefs and demand his removal from the province.

The chiefs' initiative was unprecedented; I have found no earlier record

---

[30] Like the 1990 coup by Lt. Luchembe that prefigured Kaunda's fall, the decisive event of the 1997 coup attempt was the takeover of the ZNBC broadcasting facility in Lusaka. On the morning of the coup, Zambians woke up to the sound of reggae on the State wavelengths and an announcement by 'Capt. Solo' that the military had taken over rule of the country. Radio being the only media with real national coverage, the whole nation knew of the coup immediately.

[31] This is partly speculation. I was told by an official source (but 'off the record') that signatures to the Chiefs' letter were collected using a vehicle from the Provincial Permanent Secretary's office.

C/O MANSA MUNICIPAL COUNCIL
P.O. Box 710001
MANSA

20th February, 1998
The President,
Mr. F.J.T. Chiluba
State House,
LUSAKA

Mr. Chiluba,

We are sending you our greetings as Chiefs based in Mansa District. We thank you very much for the manner in which you have brought us back respect as leaders of the land. But what is happening with the people you have appointed to assist you in leading the people?

We have a complaint against the Deputy Minister for Luapula Province.

1. He has no respect for us as Chiefs of the people in Mansa District.
2. He even wants to put some Chiefs in prison based on charges which are not justified.
3. He is in continual conflict with his colleagues. The situation is very bad and can even retard development in the Province.
4. He has developed a bad habit of separating the Chiefs of the Lunda in the Luapula valley from the Aushi Chiefs in Mansa District.

To prove this:-

a. Whenever the First Lady [the President's wife] is making a trip to Luapula in connection with her work for the Hope Foundation, he always makes sure the programme only includes the Valley leaving us with no visitations![1]
b. If we have erred him why can't he just tell us openly?

Our son, what we want is for the Minister to be removed from here so that we can be relieved.

May the Lord almighty bless you on your way.
We remain patiently awaiting your response.

We, the Chiefs [of Mansa District.][2]

cc. Office of the President, MANSA

---

[1] The Hope Foundation has clear clientelist features and might be seen as a 'female' version of the President's K10 billion discretionary fund which is used to support Chiluba's important client groups, like evangelical Churches. Hope Foundation is run by Mrs Vera Chiluba, the President's wife. Its mandate is to distribute goods and fund small charity projects among the poor.
[2] The letter was signed by the following Aushi leaders: Senior Chief Milambo, Chief Mabumba, Hon. Chief Mibenge, Sub-Chief Kale, Chief Chimese, Sabukifu [Sub-Chief] Chamawabuseba, Sub-Chief Nsonga, Chief Kasoma Lwela, Sub Chief Kundamfumu, Chief Matanda, Sub-Chief Kapwepwe, Sub-Chief Chansa, Sub-Chief Mabo, Chief Kalasa Lukangaba, Chieftainess Kalaba, Chief Chisunka

Figure 23: a translation of the original letter in ciBemba. Thanks to Chris Chambula of Credit Management Services in Mansa for assistance with the translation.

of a comparable, coordinated intervention of Zambian chiefs into party politics. It also represents the unique and similarly unprecedented marriage of interests of a senior administrator and traditional leaders against a high-ranking politician in the ruling party. Neither the alliance between chiefs and civil servant, nor the confrontation with a high-ranking politician would have been conceivable under the Second Republic. Nevertheless, the manoeuvre was not successful. Despite the chiefs' complaints, the deputy minister remained in office (although he was eventually promoted – recalled? – to Lusaka, ultimately rising to a full ministerial portfolio).

Indeed, the specific outcomes of this episode for its dramatis personae are far less significant than what it reveals about nascent chiefly agency in the Third Republic. The pivotal line in the chiefs' letter is the one demanding the minister's removal. Portentously, it begins with the words *we mwana wesu,* literally: you, our child. The familiar form of address is in stark contrast to the high formality with which the letter begins. The chiefs are telling the president that the while they recognize him as the secular ruler of the nation, in terms of their shared tradition, he is still their son and their subject.

Taken in such an explicitly political context, this rhetorical inversion of the role of ruler and subject represents a forceful gambit. But how was this strategy grounded in Aushi society? Where is the community in these episodes? Chief Mabumba sought to carve out political space on two fronts – in a tug-of-war with Chief Milambo within the insular sphere of Aushi royalty, and vis-à-vis the state in his crusade against the president's nephew. Both were bold moves, yet neither involved any apparent consultation with his subjects. Certainly, organizing the demanded a minimum of community support. Mabumba's subjects contributed tributary labour for clearing the grounds and building the visitors' shelter. Yet Chief Mabumba hardly conferred outside the circle of his immediate family and advisors concerning his campaigns for the senior chieftaincy or against the Deputy Minister. These initial strategic moves towards a larger political arena did not involve, or require, the mobilization of community support.

With the advantage of hindsight these episodes signal a clear shift in the politics of chieftaincy in Zambia. Taken together, these manoeuvres in remote Aushiland represented a determined political strategy to expand the scope of chiefly agency and speak of a strategic attempt to consolidate and expand the political capital one can now see at work in the growing assertiveness of the royal foundations in national politics. Through his inspired gambits, Chief Mabumba emerged, temporarily, as the primus motor of the Aushi chiefs. For several years to come, he oversaw a Makumba ceremony in his chiefdom. But his rise to pre-eminence was short-lived. In 2003, the *Makumba-wa-Chilololo-Chalwe-Bulimi* ceremony was abolished in favour of a new annual *Chabuka* festival, to be hosted by Chief Matanda, commemorating the original crossing of the Aushi people over the Luapula river in their eastern migration from their ancestral home in Kola. This decision could be seen as a reprimand to Chief Mabumba, and perhaps it is, but what is most striking about this cultural revolution in Bwaushi is not its implications for the political intrigues within the Aushi royal clan, but the fact that the Chabuka festival is no longer a strictly local affair; it is sponsored and overseen by a Lusaka-based committee of urban, professional Aushi. Indeed, the Lusaka-based Chabuka Committee is the decisive player behind the decision to redefine the site and substance of the main Aushi cultural 'tradition.' When asked about their decision to do away with the Makumba festival, a committee member explained that the main reason was their estrangement from the mystical underpinnings of Makumba – 'We can't see this spirit, but at least we can see the place where our ancestors forded the Luapula!'[32]

Reflections: Chiefs, State and Democracy

The 'deregulated' political and economic environment of the Third Republic has facilitated the emergence of a chiefly mode of national political agency. On one level, the overall trend in Zambia is very similar to that in South Africa where 'the lack of an active state leaves room for the re-invention of tradition and... the urban political elite might find this re-invention suitable to its own political agenda', and where 'traditional leaders are forced by circumstance to adapt to a rapidly changing social, economic and cultural environment' (Koelble 2005: 4). Compared with South African chiefs, however, Zambian traditional leaders confront these new political opportunities with a far weaker institutional base.

Zambian chiefs can claim a popular mandate, assembled and maintained in a rhetorical registry quite distinct from that of the middle-class elite's

---

[32] In actuality the road to Chief Matanda's area is so bad that the Chabuka festival has been held near the Mansa Research Station in Chief Chimese's area (interview with Chabuka Committee member Gladys Chipanta, Lusaka, 24 August 2005).

political projects. As new political opportunities arise, to what ends will the chiefs exercise their mandate? The political options, and tensions, are not those posed in the South African context, i.e., of a standoff between tradition and democracy at the local government level (the 'two bulls in the stall' dilemma). Growth in the chiefs' political agency cannot threaten a 'representative' form of local government that, for all intents and purposes, does not exist. Rather, the chiefs' entry into elite political society has direct implications for the way that the representation of popular interests in government is conceptualized and institutionalized.

Ironically, perhaps, many Zambians perceive the current political juncture is as a moment of democratic opportunity: non-partisan forces – the church, the women's movement and lawyers – acting in the name of a greater good have banded together as the Oasis Forum to defend the constitution, and resist the abuse of executive power. Chiefs have been cautious in their relation to this unusual alliance. The elite chiefs, especially, have made evident overtures toward the Oasis Forum and its agenda. Chiefs have participated in numerous Human Rights workshops organized by Women for Change (a key player in the Oasis Forum), and have signed onto human rights declarations that articulate a 'universalized' version of customary law. Chiefs have come out vocally for condom use (something the socially progressive Catholic Church refuses to do) and are active participants in the HIV/Aids campaigns of many NGOs. They call for an end to corruption and, most significantly, have joined the Oasis Forum and the independent *Post* newspaper in vocally challenging the current administration. Witness the following statements from the *Post* in June 2005:

> According to Chief Ishindi of Northwest Province 'Politicians should not hijack the entire political system because chieftaincy is also about politics. The same issues of food, water and land which politicians talked about are the same issues chiefs deal with in the villages. Therefore this system of saying chiefs should only concentrate on culture is not good. Chiefs should also be autonomous, we should have the same powers that chiefs in other nations have such as Nigeria, Ghana and South Africa,' (quoted in Kabwela 2005).
>
> Chief Nzamane, who is also Eastern Province Royal Foundation chairperson, said he was disturbed with President Mwanawasa's reaction to his complaint about Eastern Province's lack of representation at cabinet level. 'I have been following the events after the statement that I released on behalf of the traditional leaders of the province with dismay,' he said. 'Those in government should realise that we are not begging for positions but it is our birth right to have representation at cabinet level and enjoy equal access to development opportunities ... [Government] must realise that the statement was out of our

resolutions endorsed by Royal Highnesses during one of our consultative meetings. So it is not my statement but a collective voice of the chiefs,' he said (quoted in Phiri 2005).

Chiefs seem poised to join the ranks of militant civil society but, on the other hand, they cannot cut themselves off from the state; their fundamental strategy is and has always been to elicit more comfortable levels of government patronage. What path will they follow?

Koelble's conclusions concerning the future role of South African chiefs suggests possible options:

> ...the door is wide open for several, quite contradictory developments to occur. On the one hand, it is entirely possible that an African version of local level democracy in which deliberations and consensus-building represents the central core of values develops. ... On the other, it is also a possibility that traditional leaders take the opportunity to establish ... 'limited sovereignty tyranny'. ... Currently, the neo-liberal framework dictated to the post-colony in general points to the later outcome as being the more likely. It is also possible that several arrangements between these two polar opposites emerge (Koelble 2005: 30-31).

For Zambians, the likelihood of 'an African version of local level democracy' seems remote, however. Unlike South Africa's ANC, neither the MMD nor its main competitors are committed to the devolution of powers to local government, nor is there a clearly articulated political demand for such a policy. Euphemisms about 'incorporating traditional authorities into local-level governing structures, in ways which take account of local cultures' (Crook & Manor 2001: 28-9) have little currency when 'local government' is itself a euphemism.

The irony of Zambian democracy is its persistent top-heaviness. The options confronting traditional leaders, and their potential allies, have little to do with the fortunes of marginalized rural society. The national, elite, activist chiefs struggle to empower the House of Chiefs, but without constitutionally recognized links to viable institutions of local government, the House of Chiefs lacks political teeth. Leverage in Lusaka *can* translate into rural investments like roads, bridges and clinics. Rural subjects *can* benefit indirectly from their chiefs' rise in political stature, but nothing indicates that greater official recognition of chiefly authority will translate into the political empowerment of their subjects and their greater influence as citizens over the basic facts of their existence.

Despite Chief Mabumba's efforts, the numerically small, economically and geographically marginal Aushi people have not managed to catapult a member of their royal clan into the national political arena. Yet, in one sense, their chieftaincy has been 'nationalized' nonetheless. While no Aushi chiefs have established a presence in the inner circles of Lusaka's cosmopolitan elite, key aspects of royal authority appear to have been co-opted by urban-based, middle-class Zambians of Aushi heritage. Where at first blush it appeared that the processes of deregulation introduced in the Third Republic have enhanced the importance of localized political agency, the contrary may in fact be true. Rather than empowering local actors and highlighting horizontal political bonds, deregulation has instead fostered a delocalization of politics, and shifted the crux of political accountability from the horizontal to the vertical plane.

Paradoxically, deregulation has augmented the political capital of the chieftaincy, but the currency of this capital has been dislocated. That is, the potential of chiefs to leverage benefits from the state can only be realized in the national arena, far from the grassroots sites where chiefly legitimacy is constituted. This is the lesson of the emerging guild of 'national' chiefs. Furthermore, when no chiefs have managed to jump scale to the national arena – as in the case of Aushi royalty – the augmented cultural capital of the chieftaincy can be wielded by a surrogate group of urban elites.

The real issue then is not of 'tradition' versus 'democracy', but of the democratic ambitions of the new urban elites who seem fully capable of subsuming the local mandates of traditional leaders under their own agendas. The few chiefs who have established themselves within the circles of the new political society are doubly privileged in terms of their political capital. They can claim a popular mandate from their subjects, while also drawing on the momentum of the new urban elites. This dual entitlement explains their uncommon political visibility in national affairs. Still, the chiefly dimension of this entitlement is extremely fragile and depends on recognition by urban-based allies. Unlike in South Africa, claims based on customary law have not succeeded in challenging the liberal tenants of the Zambian constitution, a clear indicator of the subsidiary status of the chiefs' 'Kingdom of Custom' (Comaroff & Comaroff 2005), relative to the statutory legal regime. It seems unlikely, then, that the amalgamated Royal Foundation of all Zambian chiefs will have much currency vis-à-vis the critical political outcomes of the coming decade – the fate of the ongoing constitutional reforms (upon which the distribution of powers within the polity hinge) and the economic policy framework (which will largely determine the extent of Zambia's sovereignty in relationship to South Africa, China and the EU). These outcomes are in the hands of the thin elite veneer of Zambia's professional classes.

There are indications that it is possible to build bridges between liberal ('rights-driven') civil society and traditional leadership. At issue is the extent to which chiefs can fulfil the task at which political parties seem to consistently fail: bringing the interests of the agrarian population into political deals and contestation. The weakness of the Oasis Forum and other urban representatives of cosmopolitan civil society is their weak links to the grassroots. By default, chiefs are considered the most 'embedded' grassroots political authority for much of the population (Crook & Manor 2001). Their participation in the liberal politics of the Oasis Forum could energize the popular basis of the progressive social forces based in Lusaka.

On the other hand, the new national royalty, absent, alienated and dislocated from their roots in the land may revert to its 'customary' conservatism. Years of penury make them susceptible to the allures of state patronage, and to the attractions of the clandestine 'privatization' of natural resources. In the end, chiefs may betray the democratic potential in the current juncture and become another elite stratum, participating with the other elites in the distribution of political spoils, and behaving like other Lusaka-based groups whose claim to membership in political society is their claim to 'know' and represent 'the people.'

6

ZAMBIA'S 2006 ELECTIONS
The ethnicization of politics?
(2007)

On 28 September 2006, Zambia went to the polls in its fourth general elections since the restoration of political pluralism in 1991. Like in the previous tripartite elections in 2001, the presidency was heatedly contested, as were parliamentary and local government seats in most constituencies. Thirteen parties participated at some level, and five fielded a presidential candidate. The ruling Movement for Multiparty Democracy (MMD) incumbent, Levy Mwanawasa, retained his seat at State House with 43 percent of the ballot, while his two main opponents, veteran firebrand Michael Sata of the Patriotic Front (PF) and newcomer Hakainde Hichilema of the United Democratic Alliance (UDA) received 29 percent and 25 percent of the vote respectively.

MMD won 73 of the 150 parliamentary seats to be filled by the ballot. It only retains control of the legislature by virtue of eight deputies appointed directly by the president. The ruling party's mandate decreased only little in comparison to its pre-election status, and it avoided the embarrassing implosion predicted by the opposition. Yet, with its razor-slim majority in the National Assembly and a minority president in State House, its legitimacy is a far cry from the three-quarters' quorum it enjoyed throughout the 1990s. Given the frequency of by-elections in Zambia (due to the high mortality of office holders), MMD's parliamentary majority is very tenuous indeed.

More critically, perhaps, MMD has been completely marginalized in the major municipal councils along the line of rail. The Patriotic Front has hegemonic control of local government institutions in Lusaka, in the influential Copperbelt towns and in Kasama in the populous Northern Province.

The UDA controls Livingstone. In principle, local political institutions are in opposition hands in all of the main population centres of the country.

Local and international monitors generally hailed the elections as free and fair, albeit not without their share of technical problems. The turnout was a respectable 71 percent and the actual polling proceeded peacefully without major incident. All in all, it would seem that basic democratic procedures are becoming routine in Zambia. This overall impression was marred by a brief flare-up of mob violence in Lusaka, the national capital, as frustrated supporters of unsuccessful presidential aspirant Michael Sata took to the streets, accusing the ruling MMD of election fraud.

The Sata Factor
Notwithstanding its failure to capture the presidency, the uncontested victor of the elections was the Patriotic Front under the leadership of 69 year-old veteran politician Michael Sata. Increasing its share to a walloping 43 seats, up from a mere two in the previous parliament, PF's success was most striking in influential urban centres where it swept both parliamentary and local government seats.

PF's explosion into the major league of Zambian politics came as a surprise to most Zambians. As little as ten months before these elections it was difficult to muster even lukewarm support for Sata among Lusaka's political illuminati. Just days before the elections, the independent and influential *Post* newspaper – considered a mouthpiece for the progressive middle-class, and definitely no friend of the ruling MMD – ran a scathing attack on Sata. For once, it seems, no one was reading the *Post*.

Sata began his career as a political lieutenant to founding president Kenneth Kaunda in the heyday of his United National Independence Party's (UNIP) 'one-party participatory democracy'. Nick-named 'King Cobra' by friends and detractors alike, Sata soon carved out a distinctive niche for himself as an aggressive, rough-mouthed muscleman, incessantly poised to attack dissidents within the ruling party. It was a role he subsequently sequelled at the elbow of Kaunda's usurper, President Frederick Chiluba of the MMD.

Swept into power in 1991 by throngs of near-ecstatic citizens fed up with Kaunda, UNIP and incessant economic decline, Chiluba espoused the rhetoric of liberalism and democracy. Once in power, however, the MMD gradually sank into a morass of corruption and abuse. In 1994, then Vice-President (and current State House incumbent) Levy Mwanawasa quit government in protest over growing corruption in the MMD. Mwanawasa's resignation was in direct reaction to a shady deal he attributed to Sata. A very personal animosity between the two men has continued to the present day.

Sata remained adamantly loyal to Chiluba almost up until the end. Chiluba was constitutionally obliged to step aside in 2001, having served the maximum two terms at State House. Sata clearly expected to be anointed as his successor. But Chiluba procrastinated in declaring his intentions, and in doing so incited popular mobilization against an alleged third-term bid by the president. The nation-wide 'Green Ribbon' campaign, spearheaded by the activist Oasis Forum (a loose alliance of all major Christian church bodies, the women's movement and the Law Association of Zambia), proved incontrovertibly that the Zambian people would not countenance another five years of Chiluba.

As time ran out, Chiluba sidelined the unpredictable Sata and identified ex-Vice-President, lawyer Levy Mwanawasa, as his heir apparent. It was a surprising and unconventional move that Chiluba lived to regret. After several years of barely concealed abuse of public assets, Chiluba needed desperately to ensure that his successor would protect him against accusations of financial impropriety. His choice of Mwanawasa demonstrated a serious failure of character assessment on Chiluba's part. Apparently he believed that Mwanawasa, estranged from MMD inner circles and who, it was rumoured, had never fully recovered from a head injury in the early 1990s, would be easy to control. As it turned out he was mistaken.

Be that as it may, Sata was visibly shaken by this unexpected turn of events and left the MMD with doors banging. He quickly formed the Patriotic Front along with Guy Scott, a member of Zambia's numerically insignificant troupe of white settlers. The hastily assembled PF did poorly in 2001, and its performance in subsequent by-elections was not impressive. On the eve of the 2006 polls PF held only two parliamentary seats, in contrast to main opposition party, the United Party for National Development (UPND)'s 49. For the first time in his political career, Sata was on the outside looking in and he didn't like it.

It is hard to link Sata to any clear ideological platform. He is known as a fixer and a hard worker. While district governor for Lusaka in the late eighties, for example, he provided affordable housing to many urban residents and achieved the Herculean feat of cleaning up a decade of accumulated rubbish on the city's streets. He can also work a crowd better than any contemporary Zambian politician. His defining trademark is gravelly, populist rant, never far from the gutter, that revels in hyperbole and political taunt. When explosives were discharged in July 2005 at Konkola Copper Mines, in connection with worker-instigated protests against a privatization scheme, Sata rushed to the scene to claim complicity in the bombing. (As a result he was arrested on sedition charges, a case that is still pending.) And on the eve of the recent elections, he praised Robert Mugabe's violent land seizures in troubled Zimbabwe, while in the same breath threatening alien (Asian and Lebanese) businessmen in Zambia with deportation.

Such brutal demagoguery is rare in Zambian political society. Yet Sata's campaign maintained a counter-intuitive upward swing as the 2006 elections approached. PF rallies pulled large, buoyant crowds wherever he spoke. Major opposition politicians like former UPND Vice-President Sakwiba Sikota and firebrand Given Lubinda defected from their mother party to join the PF bandwagon. The diplomatic corps was nervous. Murmurings about the 'Zambian Mugabe' circulated in the capital with increasing anxiety.

Come election day, PF went to the polls confident of victory. Amazed citizens stayed glued to their radios and TVs as the preliminary count pointed to a PF landslide. Early returns from urban constituencies had Sata leading Mwanawasa almost 2:1. In the final count, PF swept Lusaka and the mining towns of the Copperbelt, and garnered substantial support in the 'Bembaphone' northeast. Elsewhere – with the exception of the Tonga-speaking Southern Province, where UDA candidates harvested all but one seat – MMD prevailed. Since the Copperbelt population is also predominantly ciBemba speaking, one might argue that PF's victory is evidence of the 'ethnicization' of Zambian politics. The fact that the UDA's electoral success was limited to one, ethnically homogenous region also lends credence to such an interpretation. Closer inspection, however suggests that the ethnic explanation may be too simplistic. I return to this point further on.

It is probably fair to say that PF's success at the polls was to a large part due to pre-election fumbles by both MMD and UPND/UDA. MMD's main liability is Mwanawasa himself. Zambians have little genuine affection for, much less fear of, 'Levy'. Once a successful Copperbelt lawyer, Mwanawasa's public persona exudes impulsiveness and arrogance, coupled with a propensity for alienating legalese. He also suffers from periodically debilitating health problems. Hot on the heels of his 2001 victory, Mwanawasa won some popular sympathy by bringing his mentor Chiluba to trial on corruption charges. He nevertheless quickly squandered this windfall popularity through inconsistent policies, nepotism and petty squabbles with civil society groups like the Oasis Forum.

Mwanawasa is also out of touch with popular demands for delimiting presidential powers and expanding socio-economic rights. After vowing to honour the recommendations of the Constitutional Review Commission he appointed in 2003, Mwanawasa distanced himself from the draft constitution they produced which, among other things, required that the President win more than half of the popular vote. Having squeaked through on a (highly contested) 29 percent plurality in 2001, Mwanawasa was understandably uneasy about his chances for re-election in 2006 under such a provision. Through filibustering and political manipulation, MMD stalled constitutional reforms with the result that the 2006 elections were held under the simple majority clause introduced by Chiluba in 1996. From the MMD's perspective this was a prudent tactical move. It is anybody's guess how Mwanawasa would have fared against Sata had the recent elections gone into a second round.

Mazoka's Ghost

PF's dramatic advance benefited directly from the collapse of the hitherto most credible opposition force, the UPND. In 2001 UPND's founding president, ex-Anglo American executive Anderson Mazoka, lost to Mwanawasa by less than two percent of the popular vote. In reality, Mazoka probably had the greater share of popular support, but was deftly out-manoeuvred by the MMD which ruthlessly exploited its control of state resources during the campaign period. (The Supreme Court ruled against UPND's petition to overturn Mwanawasa's 2001 election on the grounds of unfair practices, yet the protracted hearings brought forward massive evidence of MMD manipulation as well as rigging by all parties.) Mazoka fell seriously ill soon after his defeat and spent much of Mwanawasa's first term of office under intensive care in South Africa. He returned to Zambia in 2005 and resumed leadership of UPND. Despite his evident frailty he succeeded in suppressing efforts to replace him by divisive factions within the party.

After Mazoka's death in May 2006 at age 63 things fell apart, and UPND split over a secession crisis that had two debilitating consequences. One, the sidelining of senior UPND stalwarts in favour of political novice Hakainde

Hichilema as party president reaffirmed popular conceptions of UPND as an ethnically grounded Tonga party. Second, the split saw the defection of popular UPND mainstays Sakwiba Sikota and Given Lubinda into an alliance with PF. Although Hichilema's 25 percent share of the presidential vote is a respectable achievement for a political unknown, UPND's share of seats in the new parliament decreased by almost two-thirds. All of its current seats are in the Tonga-speaking constituencies of the Southern Province.

Primordialism Resurgent?

All in all, the technical quality of these elections was a clear improvement on previous multiparty polls. This time around, the MMD government made a concerted effort to allay accusations of pre-election machinations. Cabinet was dissolved well ahead of time and Ministers were not permitted to use government resources for their campaigns. There are around twenty court petitions pending in contest of constituency-level results but given the technical complexities involved in an exercise of this scope, this is hardly unusual.

Did these elections signal the ethnicization of Zambian politics? 'Tribalism' is a register generally eschewed in public political discourse in Zambia. The fact that UNIP managed to rule for 27 years with few or no signs of ethnic tension is still considered an unmitigated virtue in Kaunda's complex political legacy. That said, some observers are convinced that ethnic identity and rivalry simmer ominously beneath the surface of Zambian politics (e.g. Posner 2005). The fact that more than half of the popular vote went to candidates with strong ties to Bemba or Tonga constituencies would appear to lend support to this claim.

While deepening political pluralism is bound to enhance the currency of many sorts of 'primordial' identities – of race, gender, religion as well as of birth-place and language – I doubt that ethnicity was a decisive factor in the electoral outcome. Both Mazoka's untimely exit and Sata's last-minute upward leap are primarily contingent as against tendential, structural factors. In some parts of the country, such as Loziland to the West and Ngoniland to the East, ethnic alignment in electoral politics seems to be, if anything, on the decline. What is indeed striking about PF's campaign was not so much its 'ethnic' character but its brash contrarianism, and the appeal of such demagogic radicalism to members of the urban underclass across ethno-linguistic boundaries. For all his populist bravado, Michael

Sata has brought real issues of concern to the urban poor into the political arena.

Given the strong role of contingent factors in the election results it is unusually hard to project far-reaching trends. It is clear, however, that a sea change of sorts is underway. In a conventionally winner-takes-all political culture, the nominal winner, the MMD, lost more than it won. Two sites of struggle emerge. One, the politically volatile urban councils, where PF has an unprecedented opportunity to institutionalize its grassroots support through improved performance in water, sanitation and housing – the main demands of the urban poor. Two, parliament itself, where a united opposition can force the government's hand on, among other things, constitutional reform.

At root, the 2006 polls should be seen as a protest election and not a retreat into primordial politics. Sata's uncanny avalanche was a clear message to the political class in general and Mwanawasa in particular. 'We wanted to rub salt in the wound', as one Sata supporter put it. Deepening social and economic disparities are generating anger and frustration. Zambians want leadership, not excuses.

7

EPILOGUE
(November 2009)

Things seem, once again, to be changing rapidly in Zambia. Not only has the rate of the kwacha against the dollar been volatile, but policies that were the cornerstone of late President Mwanawasa's rule seem far less certain under the watch of President Rupiah Banda. When I last enjoyed a longer stay in Lusaka, in 2004, it appeared as if 'civil society', in the form of the Oasis Forum, had the late Levy Mwanawasa's government in full court press over constitutional reform. Today, one year into Rupiah Banda's tenure of office, government is once again being run with impunity by the MMD National Executive Committee. Of the Oasis Forum's core leadership, the Law Association has gone into hibernation, the Catholic clergy has been hobbled by a savage government attack on its integrity, and the women's movement has publicly distanced itself from the *Post*, once the Forum's truest ally.

Politics aside, the constant state of flux reaffirms Zambia's status as a society in transition. What is less clear, however, is the nature, and destination, of this transition. On one count, Zambian society is still coming to terms with its colonial past. The continuing legacy of colonial-era structures, including the bulk of the Constitution and of the laws on the books, affects the relationship between citizens and rulers in countless ways. On the other hand, Zambia also continues to struggle with the long-term effects of the command economy established under UNIP, with a steeply skewed dependence on copper, and with deep-rooted reliance on foreign aid. Indeed, we might say that Zambia is caught up in a 'transition cycle': few if any of these various processes of change are moving forward in a clear or systematic way. On the contrary, many grievous legacies of Zambia's past continue to haunt leaders and citizens alike.

This collection of essays has sought to highlight such elements of continuity amidst rapid and often profound change. The point is not to breed fatalism, but to encourage more open and determined engagement with the ballast of the past. The main focus has been on identifying persistent obstacles associated with integrating rural producers into the national economy. Economic citizenship is an end in itself, but it also helps secure democratic participation in defining the means and ends of the nation's development. Rural farmers have generally lost out on both counts. Anyone with even a passing familiarity with the realities of rural life in Zambia today will recognize that, for all its much-touted 'potential', agriculture remains a back-breaking, unrewarding and uncertain livelihood for the vast majority of peasant farmers, much as it was at independence forty-five years ago. Chiefs and MPs offer an official channel for redress but, increasingly, both politicians and traditional leaders often seem distracted by concerns related more to their own, rather than their constituencies', fortunes.

The failure of agrarian transformation is not completely due to a lack of political commitment to rural development. The UNIP government invested substantially in the promotion of hybrid maize cultivation, and in the co-operative marketing institutions commensurate to that task. Over the past decade, the Ministry of Agriculture (in its various incarnations) has received about 6% of the national budget. Of this, according to a World Bank study, about 40% has been ploughed into the government's Fertilizer Support Programme (World Bank 2007). Indeed, between the 2002/03 and 2008/09 growing seasons, GRZ budgeted more than 1,360 billion kwacha (almost US$300 million at current exchange rates) to the FSP, involving 422,000 metric tons of fertilizer and reaching a cumulative estimate of 1.5 million farmers (ZACF 2009).

I do not have sufficient data to compare the current level of expenditure with that invested in agricultural subsidies in the 1980s, but one thing is relatively clear. The neoliberal MMD government's resort to UNIP-style input subsidies at the core of its agricultural policy suggests that this particular policy instrument is rather independent of the ideological commitments of the political party in power.

Despite two decades of deregulated development in Zambia, then, agrarian transformation is still stalled at square one for much of the country. The fundamental problems of low productivity and ineffective demand persist, as does the main policy instrument – input subsidies – with which government seeks to redress them. What is most disheartening about this pattern is the persistent mismatch between expenditure and results over the decades. Why isn't the recipe working?

As noted, the current Fertilizer Support Programme (FSP) represents the main public intervention in rural development. Given the importance of the rural vote, it is thus not surprising that over its seven years of operation, the FSP has been a constant target of criticism. Recently, an assessment sponsored by the Danish NGO MS-Zambia accused the FSP of being 'rocked with corruption and [...] highly political.' The FSP was deemed 'extremely expensive and bureaucratic,' and its author, one Michael Muleba (2008), even suggested that the programme was 'taking resources away from the very farmers it intended to serve.' Muleba identified a number of serious problems in the FSP related to the timing of input delivery (adversely affecting yields), to the high cost of the subsidy to government, and to the lack of effective safeguards against corrupt practices in the distribution of subsidized inputs.

A parallel review of the FSP (ZACF 2009), led by senior civil servants in the Ministry of Agriculture and Cooperatives, largely concurs with this view. While corruption is not mentioned directly, the MACO review does note concerns with 'poor targeting of farmers/beneficiaries.' Other stakeholder concerns noted in the report include:

- Delays in input distribution;
- Poor fertilizer use efficiency among targeted farmers;
- Inconsistency of policy implementation, especially in reversal of plans to reduce the subsidy level, and to stimulate agro-dealer development;
- [Negative] FSP impact on private sector participation;
- Long-term concerns about the FSP sustainability;
- Poor monitoring of program effects making it difficult to sort out program achievements against objectives.

I find it striking that this list would have been just as valid in 1991 or, concerns about market distortions notwithstanding, in 1975. How should we interpret this? The government is prepared to invest an average of US$43 million a year on a strategy which has failed consistently to produce tangible results since its inception more than 40 years ago!

The most prevalent explanation for this apparent paradox is to see to this pattern of unproductive public expenditure as a prime example of 'neopatrimonial patronage' (Eberlei et al 2005). The party in power is not genuinely concerned, it is claimed, with rural livelihoods, food security or agricultural modernization. Rather, these funds are meant solely to secure the political support of rural constituencies. While there is much truth in this, the neopatrimonial explanation has worn increasingly thin in the age of multi-party politics. As electoral results have become increasingly unpredictable, it can hardly be politically rational to provide blanket patronage to constituencies which may well support the opposition. A more carefully targeted approach would no doubt produce better results. And if, as is widely claimed, the lion's share of the limited supplies of subsidized fertilizer ends up in the hands of ruling party loyalists, this could easily prove politically damaging for those in power. Why re-elect a candidate whose friends grab the few meagre benefits government has earmarked for the folk in the village?

At the heart of a neopatrimonial relationship is the premise of trust: the people's trust in government, and the government's trust in the people, are preconditions for the definitive neopatrimonial transaction of exchanging money (or mealie meal, beer and t-shirts) for votes. The mainstream view on patronage in post-colonial politics makes the mistake of assuming a basis of trust between government and citizens. It is more realistic to doubt that sufficient trust exists between the parties for neopatrimonial transactions to predominate. I suspect, rather, that there is another logic at work, even more fundamental than the political rationality of patronage: that of exclusion.

The persistent failure of government programmes for agrarian transformation is not primarily the result of 'hidden agendas', they are not the outcome of secret policies worked out in smoke-filled rooms. Nor can they be attributed to the dishonesty of corrupt individuals, although corruption clearly diverts a substantial part of the public investments away from its intended beneficiaries. Rather, persistent agrarian stagnation might be better understood as an unintended consequence of a structural logic of social exclusion that divides Zambian society with ruthless consistency. Smallholder, semi-subsistence farmers cannot labour their way out of economic marginalization because they are systematically barred access to the means of success – a decent education, reliable health care, the most productive land, the best roads and bridges, affordable credit, access to lucrative markets, and so on. On the surface, this seems to be a function of their place of residence. But a spatial explanation alone cannot account for the way that 'business as usual' systematically undermines the economic and political citizenship of rural and urban poor alike.

Under colonial rule, social exclusion was a function of race, with all significant privileges and benefits reserved for Europeans. Since Independence, however, the polarization of Zambian society into haves and have-nots has inevitably played out according to a different, non-racial logic. Compared with the nakedness of distinctions based on race, the mechanisms of post-colonial exclusion are harder to pin down. If you grew up struggling against blatant racial stereotypes, it can be difficult to see how exclusion now takes more subtle forms. On the one hand, those less successful can be labelled lazy, lacking in ambition, or morally flawed – drunkards and polygamists. One can even hear well-to-do individuals invoke the feudal residues of 'commoner' and 'slave' to explain others' economic misfortunes. On the other hand, the emphasis on individual liberty and welfare in the Third Republic – a stark departure from the explicit collectivism of Kaunda's Humanism – encourages one to believe that what is good for him-or herself, is also morally justified.

To characterize such attitudes as 'elitist' is accurate, but tautological. Few Zambians achieve sustainable prosperity; by definition, any well-to-do Zambian belongs to an elite stratum of society. The story of how postcolonial boundaries between the elite and the masses have been drawn, maintained and patrolled has yet to be told in full, and requires further study. I would suggest that the boundaries are at once moral and political. Moral, in that they relate to how the notion of *entitlement* is negotiated in Zambian society (Sen 1981).[1]

Why do some people seem to deserve privileges and benefits that others are routinely denied? Clearly, the understanding of entitlement is in flux. This moral transformation has both reflected and fuelled an unprecedented process of social stratification, polarization and class formation in Zambia. In times of crisis, uncertainty and relative austerity, Zambians are obliged to make subtle distinctions about the extent of our responsibilities to rela-

---

[1] Amartya Sen deployed the concept of entitlement to explain periodic famine. Sen claimed that famines cannot be explained by a shortage of food, but by the victims' lack of entitlements to food. Sen's understanding of entitlement was strictly economic, and could only be expressed through a market relation. Following Devereux (2001) and others, I am using entitlements much more broadly, in relationship to claims that people can assert in a wide range of situations, also in non-market settings.

tives, friends and strangers. At the same time, technology has devoured the physical distances across which entitlements can be claimed. While the 'common good' was once be limited to lineage kin, neighbours or a cohort of age-mates, people are now be asked to demonstrate solidarity with people they are never even likely to meet.

It is no wonder, then, that the contours of entitlement are indistinct and open to constant renegotiation. In the course of these negotiations, any given trait or marker can provide respite from unwanted claims – varieties of faith, accent, dress or politics can provide the wedge that draws an invisible but effective border of exclusion.

These borders and the distinctions that define them are, of course, profoundly political. They determine the pattern of resource allocation in society – why the extension officer's nephew is 'entitled' to an extra allotment of subsidized fertilizer, and a minister never pays for his own drinks. But the fact that the popular understanding of entitlements is in flux also means that there is room to introduce new understandings – definitions that cater to more flexible, and just, mechanisms of inclusion.

Boundaries of exclusion are also political in the sense that they can only be redrawn through political action. What is required is enlightened political leadership, by both those in power and those seeking it. A focused programme of input subsidies, or other measures in support of improved rural livelihoods, could contribute to breaking the vicious circle of poverty and exclusion in the Zambian countryside – but only if those both claiming and distributing these benefits can agree upon just terms of entitlement. Defining what is just, of course, is part of the struggle. The better-off feel entitled to their prosperity; most believe that what they have is hardly enough. Clearly, it is time to rethink the relationship between obligation and entitlement.

This is not a new idea. President Kaunda always put 'love' at the centre of his doctrines. Clergymen, especially the Catholics, have preached 'charity' for centuries. But sermons and speeches have not carried the day. When the Oasis Forum pressed the late President Mwanawasa to enact a 'people-driven' constitution, they sincerely meant well. But even the Oasis Forum, largely led by well-educated, middle-class, city-dwelling men and women of tremendous character, moral vision and courage, could not create space for the excluded to be heard (Gould 2009). The Oasis Forum, for all its progressive virtues, spoke with the voice of a lawyer, a bishop or a manager. For all its unprecedented success in demonstrating that civil society can make its presence felt, the Forum failed to resurrect the populist enthusiasm that brought MMD to power almost twenty years ago. Granted, this was never the Forum's mission. But its isolation from 'the people' on whose behalf it spoke perhaps explains the ease with which the current administration has neutralized its influence.

And at the end of the day, the problem isn't really about the constitution – vital as it is. Even the Second Republic Constitution guaranteed individual citizens with a broad range of liberal rights. Some of these were undermined at times through derogation clauses, or due to an inconsistent judiciary, but at the bottom line, even under UNIP's authoritarianism, all Zambians were constitutionally guaranteed equal rights.[2] A constitution can always be improved, but even limited rights will not be realised if only some of the citizenry is truly entitled, in practice, to a fair share of the national wealth.

How possible would it be to establish a new politics of inclusion, where no one claims the right to speak for others, and everyone learns to listen to, and understand excluded voices? Not easy, I'm afraid. For this to happen it would require a new political awakening. This, in turn, would require wide recognition of the threat to social order incumbent in deepening injustice. Zambia's most precious asset is not its copper, but its peace and its sense of unity. The nation's oneness can only be abused so long by a politics of exclusion before unity gives way to divisiveness and conflict. The best safeguard against this disastrous eventuality is to ensure that the entitlements which are constitutionally guaranteed to every Zambian are, in fact, respected.

---

[2] A derogation clause allows someone, usually the President, to temporarily suspend basic rights, for example, in time of war or other emergency situations.

BIBLIOGRAPHY

Allen, J. M. S. (1988) 'Survival in Mabumba: putting agriculture into its context,' *ARPT (Luapula) Labour Study Report No. 1.* Mansa: Adaptive Research Planning Team.

Bates, Robert (1976) *Rural responses to industrialization. A study of village Zambia.* New Haven: Yale University Press.

Baylies, C. and M. Szeftel (1984) 'Elections in the one-party state' in Gertzel, C. et al. (eds.) The *Dynamics of the one-party State in Zambia.* Manchester: Manchester University Press.

Booth, D. et al (1993) *Social, cultural and economic change in contemporary Tanzania: A people-oriented focus.* Stockholm: Sida.

Brelsford, W. (1947) *Copperbelt markets.* Lusaka: Government Printers.

Burnell, Peter (2001) 'The Party system and party politics in Zambia: Continuities Past, Present and Future'. *African Affairs,* vol. 100.

Chanda, Donald, ed., (n.d.) *Democracy in Zambia. Key speeches of President Chiluba 1991/92.* Kitwe: African Trust Press

Chanock, Martin (1985) *Law, custom and social order. The colonial experience in Malawi and Zambia.* Cambridge: Cambridge University Press.

Chifuwe, Sheikh (2003) ' "No good governance without traditional rulers" ', *The Post* (Lusaka) 25 May.

Chinyanta, M. & C. J. Chiwale (1989) *Mutomboko ceremony and the Lunda-Kazembe dynasty.* Lusaka: Kenneth Kaunda Foundation.

Chiwele, D.K. et al (1997) *Private sector response to agricultural marketing liberalization in Zambia: a case study of Eastern Province maize markets.* Uppsala: Scandinavian Institute of African Studies.

Cliffe, L. (1978) 'Labour migration and peasant differentiation: Zambian experiences,' *Journal of Peasant Studies* 5(3): 327-8.

Coalition of Traditional Leaders [of South Africa] (2002) cited in Koelble 2005: 26.

Comaroff, John & Jean Comaroff (2005) 'Reflections on liberalism, policulturalism & ID-ology. Citizenship & difference in South Africa' in S. L. Robins (ed.), *Limits to liberation after apartheid. Citizenship, governance and culture.* Oxford: James Currey: 33-56.

Corrigan, Philip & Derek Sayers (1991 [1985]) *The great arch. English state formation as cultural revolution.* Oxford: Basil Blackwell.

Craig, J. (2000) 'Evaluating privatization in Zambia: A tale of two processes', *Review of African Political Economy,* 85: 357-66.

Crook, Richard & James Manor (2001) *Local governance and decentralization in Zambia* (mimeo)

Cunnison, Ian (1959) *The Luapula peoples of Northern Rhodesia: custom and history in tribal politics.* Manchester: Manchester University Press.

D'Engelbronner-Kolff, F.M. et al, eds., (1998) *Traditional authority and democracy in southern Africa.* Windhoek: New Namibia Books.

Eberlei, W., P. Meyns, & F. Mutesa (2005) *Poverty reduction in a political trap? The PRS process and neopatrimonialism in Zambia.* Lusaka: University of Zambia Press.

EIU (Economist Intelligence Unit) (1996) *Country Profile: Zambia 1996-7.* London

Ellis, Frank (1988) *Peasant economics. Farm households and agrarian development.* Cambridge: Cambridge University Press.

Etounga-Manguelle, D. (2000) 'Does Africa need a cultural adjustment program?' in L. Harrison & S. Huntington (eds.), *Culture matters. How values shape human progress.* New York: Basic Books: 65-77.

Gatter, Philip (1988) *Indigenous farming systems information in Mabumba.* Mansa: Adaptive Research Planning Team (mimeo).

Gatter, Philip (1990) 'Indigenous and institutional thought in the practice of rural development: a study of an Ushi chiefdom in Luapula, Zambia' (unpublished doctoral thesis, University of London, School of Oriental and African Studies).

Geisler, Gisela (1992) 'Who is losing out? Structural Adjustment, gender and the agricultural sector in Zambia,' *Journal of Modern African Studies* 30(1): 113-39.

Gordon, David (2004) 'The cultural politics of a traditional ceremony: Mutomboko and the performance of history on the Luapula (Zambia)', *Comparative Studies in Society and History:* 63-83.

Gould, Jeremy (1997) *Localizing modernity: Action, interests and association in rural Zambia.* Helsinki: Finish Anthropological Society 1997.

Gould, J. (2002) 'Contesting democracy: The 1996 Zambian elections' in M. Cowen and L. Laakso (eds), *Multi-party elections in Africa.* Oxford: James Currey.

Gould, Jeremy (2006) 'Strong bar, weak state. Lawyers, liberalism and state formation in Zambia', *Development and Change* 37(4): 921-41.

Gould, Jeremy (2009) 'Subsidiary sovereignty and the constitution of political space in Zambia', in J-B Gewald, M. Hinfelaar & G. Macola (eds), *One Zambia, many histories. Towards a post-colonial history of Zambia.* Lusaka: Lembani Trust: 275-93.

Government of the Republic of Zambia (2005), *Interim report of the Constitution Review Commission.* www.crc.org.zm (consulted 29 June 2005).

Haddad, L. *et al* (1995) 'The gender dimensions of economic adjustment policies: Potential interactions and evidence to date,' *World Development* 23(6): 881-96.

Hobsbawm, Eric & Terence Ranger (1983) *The invention of tradition.* Cambridge: Cambridge University Press.

Hutchinson, P. (1972) 'The Climate of the Mansa Area,' in R. A. Pullan (ed.), *Mansa: Zambia Geographical Association Conference Handbook* 1. Lusaka: Zambia Geographical Association.

Kabwela, Chansa (2005) 'Chiefs shouldn't be used as rubber stamps – Ishindi', *The Post,* 2 June.

Kapambwe, Stephen (2004) 'Departure of Kalonga Gawa Undi marks end of an era', *Times of Zambia*, December 13-20 (consulted 11 June 2005).

Kay, George (1964) *Chief Kalaba's village*. Lusaka: Rhodes-Livingstone Institute [Rhodes-Livingstone Papers no. 35].

Koelble, Thomas (2005) 'Democracy, traditional leadership and the international economy in South Africa,' *CSSR Working Paper No. 114*. Cape Town: Centre for Social Science Research.

Kokwe, Misael (1991) *The role of dambos in agricultural development in Zambia*. London: International Institute for Environment and Development.

Larmer, Miles (2005) 'Reaction and resistance to neo-liberalism in Zambia', *Review of African Political Economy*,103: 29-45.

Labrecque, E. (1982 [1931/34]) *Beliefs and religious practices of the Bemba and neighbouring tribes*, trans. by Patrick Boyd. Chinsali: Ilondola Language Centre.

Land Resources Study (1975) 'Land resources of the Northern and Luapula Provinces, Zambia: A reconnaissance assessment.' *Land Resources Study* 19 (Surbiton: Ministry of Overseas Development. Land Resources Division 1975), Vol. 2.

Lewis, Paul (2009) 'Shifting legitimacy. The trials of Frederick Chiluba' in Lutz, Ellen L. and C. Reiger, eds., *Prosecuting Heads of State*. Cambridge: Cambridge University Press.

Lindgren, Björn (2002) *The politics of Ndebele ethnicity. Origins, nationality, and gender in Southern Zimbabwe*. Uppsala: Department of Cultural Anthropology and Ethnology.

LRDP (n.d.), *Luapula rural development programme: Mid-term evaluation* (mimeo).

Mamdani, Mahmood (1996) *Citizen and subject. Contemporary Africa and the legacy of late colonialism*. Princeton University Press.

Mbembe, Achillle (2001) *On the postcolony*. University of California Press.

Mickels, Gun (1997) *Maize, markets and livelihoods. State intervention and agrarian change in Luapula Province, Zambia, 1950-1995*. Helsinki: Institute of Development Studies.

Miracle, Marvin (1962) 'African markets and trade in the Copperbelt,' in P. Bohannan & G. Dalton (eds.), *Markets in Africa*. Evanston Northwestern University Press.

Moore, H. L. and M. Vaughan (1994, *Cutting down trees. Gender, nutrition, and agricultural change in the Northern Province of Zambia 1890-1990*. London/Lusaka: Heinemann/Currey/University of Zambia Press.

Moore, S. F. (1994), *No condition is permanent. The social dynamics of agrarian change in sub-Saharan Africa*. University of Wisconsin Press.

Muleba, Michael (2008) 'Fertilizer support is a subsidy disaster', MS Zambia Newsletter, October 2008 <http://www.ms.dk/sw107185.asp>

Mulenga, C. (1993) 'Structural Adjustment and the rural-urban gap', IAS Working Paper, vol 1, no 4. Lusaka: Institute for African Studies.

Munro, W. (2001) 'The political consequences of local electoral systems.' *Comparative Politics*, 34, 2: 259-313.

Musambachime, M. (1981) 'Development and growth of the fishing industry in Mweru-Luapula 1920-64' (unpublished doctoral dissertation, University of Wisconsin-Madison).

Musonda, Gershom, ed. (1997) *Traditional leadership and democracy in Zambia*, Zambia Independent Monitoring Team, Konrad Adenauer Stiftung (Lusaka, Harare) .

Mutukwa, M. K. N. & O. Saasa (1995) 'The Structural Adjustment program in Zambia: Reflections from the private sector,' in K. Kapoor (ed.), *Africa's experience with structural adjustment* (World Bank Discussion Paper 288).

Ntsebeza, Lungisile (1999) 'Democratization and traditional authorities in the new South Africa,' *Comparative Studies of South Asia, Africa and the Middle East*, Vol. XIX No. 1: 83-93.

Oomen, Barbara (2000) '"We must now go back to our history"; Retraditionalization in a Northern Province chieftaincy', *African Studies* 59/1.

Philpot, Roy (1936) 'Makumba: The Ushi tribal god,' *Journal of the Royal Anthropological Society*, Vol. LXVI: 189-208.

Phiri, B.J. (2002) *Democratisation in Zambia: The 2001 tripartite elections*. Africa Institute Occasional Paper, no 67. Pretoria: Africa Institute of South Africa.

Phiri, Brighton (2003) 'Royal Foundation has endorsed chiefs' appointments on Constitutional Review Commission', *The Post* (Lusaka), 26 April.

Phiri, Brighton (2005) 'Levy's reaction to chiefs' cry disappoints Chief Nzamane', *The Post*, 5 June.

Pletcher, J. (2000) 'The Politics of Liberalizing Zambia's Maize Markets', *World Development*, 28 (1): 129-142.

Posner, Daniel N. (2005) *Institutions and ethnic politics in Africa*. Cambridge: Cambridge University Press.

Pottier, J. (1988) *Migrants no more. Settlement and survival in Mambwe villages, Zambia*. Manchester: Manchester University Press.

Quick, Stephen A. (1975) 'Bureaucracy and rural socialism: the Zambian experience' (unpublished doctoral dissertation, Stanford University).

Rakner, L. (2003) *Political and economic liberalisation in Zambia 1991-2001*. Uppsala: Nordic Africa Institute.

Sano, Hans-Otto (1989) 'From labour reserve to maize reserve. The maize boom in the Northern province in Zambia', *CDR Working Paper 89.3*. Copenhagen: Centre for Development Research.

Sen, Amartya (1981) *Poverty and famines: An essay on entitlement and deprivation*. Oxford: Clarendon Press.

Seshamani, V. (1998) 'The impact of market liberalisation on food security in Zambia', *Food Policy* 23 (6): 539-551.

Sichone, O. & B. Chikulo, eds. (1996) *Democracy in Zambia; Challenges for the Third Republic*. Harare: SAPES Books.

Simutanyi, Neo (2005) 'Chiefs and politics' *The Post* (consulted 8 January 2005)

'Traditional Leaders in Zambia: A Weapon against AIDS',
http://www.abtassociates.com/Page.cfm?PageID=12626 (consulted 10/06/ 2005).

Trapnell, Colin (1953) *The soils, vegetation and agriculture of North-Eastern Rhodesia: Report of the ecological survey.* Lusaka: Government Printer.

Van Binsbergen, Wim (1981) *Religious change in Zambia. Exploratory studies.* London: Kegan Paul.

Van Binsbergen, Wim (1987) 'Chiefs and the state in independent Zambia : exploring the Zambian national press', *Journal of Legal Pluralism,* n 25-26: 139-201. http://www.shikanda.net/ethnicity/chiefs.htm.

Van de Walle, N. & D. Chiwele (1994) *Economic reform and democratization in Zambia'* Michigan State University Democratic Governance Working Paper, No. 9.

Van Donge, Jan Kees (1998) 'Reflections on donors, opposition and popular will in the 1996 Zambian general elections', *Journal of Modern African Studies,* vol. 36, no. 1.

Verbeek, Léon (1990) *Le monde des esprits au sud-est du Shaba et au nord de la Zambie. Recueil de textes oraux précédés d'une introduction.* Roma: Libreria Ateneo Salesiano.

World Bank (2009) *World Development Report 2009.* World Bank: Washington, DC.

Wright, Marcia (n.d.) 'Legitimacy and democratization in Northernmost Zambia, 1950-1960' (mimeo).

Zambian Agricultural Consultative Forum (2009) *Report on Proposed Reforms for the Zambian Fertilizer Support Programme*
<http://www.acf.org.zm/pdf/FSP/Proposed%20Reforms%20-
%20Fertilizer%20Support%20Programme%20for%20Stakeholder%20Review.pdf>

www.ingramcontent.com/pod-product-compliance
Lightning Source LLC
Chambersburg PA
CBHW021830020426
42334CB00014B/566